Field Manual of Diseases on Trees and Shrubs

R. Kenneth Horst

Field Manual of Diseases on Trees and Shrubs

Springer

R. Kenneth Horst
Plant Pathology and
Plant Microbe Biology
Cornell University
Ithaca, NY, USA

ISBN 978-94-007-5979-4 ISBN 978-94-007-5980-0 (eBook)
DOI 10.1007/978-94-007-5980-0
Springer Dordrecht Heidelberg New York London

Library of Congress Control Number: 2013935125

Printed on acid-free paper

Springer is part of Springer Science+Business Media (www.springer.com)

Preface

The information in this manual is taken in large part from the records of the Plant Disease Survey as given in the *Plant Disease Reporter*, *Plant Disease*, and from the *Index of Plant Diseases in the United States, Agriculture Handbook 165*, U.S. Department of Agriculture. There are a great many records that bring in new reports every day, and any such list is out of date by the time it is published, and is sure to be far from complete. This check list is offered as a helpful guide, a foundation on which to build. It is by no means the last word.

Hosts are listed alphabetically by common names except where the scientific names mean less confusion. The diseases are those mostly reported from North, Central, and South America.

Contents

Part I

Trees and Shrubs 3

Part II

Disease Control and Management 13

Part III

Host Plants and Their Diseases 19

A 21

B 37

C 47

D 65

E 71

F 77

G 83

H 89

I 101

J 105

K 109

L 111

M 117

N 127

O 129

P 133

Q 155

R 157

S 165

T 175

U 179

V 181

W 183

Y 189

Z 191

Part IV

List of Land-Grant Institutions and Agricultural Experiment Stations in the United States 195

Glossary 199

Part I

Trees and Shrubs

Plate 1 Lethal Yellowing on Coconut Palm caused by a Phytoplasma pathogen

R.K. Horst, *Field Manual of Diseases on Trees and Shrubs*,
DOI 10.1007/978-94-007-5980-0_1, © Springer Science+Business Media Dordrecht 2013

Plate 2 Cedar-Apple Rust on Apple caused by *Gymnosporangium juniperi-virginianae*

Plate 3 Cedar-Apple Rust on Cedar caused by *Gymnosporangium juniperi*

Plate 4 Mosaic on Rose caused by Prunus Necrotic Ringspot Virus

Plate 5 Phyllody on Hydrangea caused by a Phytoplasma Pathogen

Plate 6 Oak Anthracnose

Plate 7 Black Knot on *Prunus* sp.

Plate 8 Botrytis Petal Spot on Magnolia

Plate 9 Azalea Leaf Gall

Plate 10 Phyllosticta Leaf Spot on Mountain-Laurel

Plate 11 Mistletoe, Common in Southern trees

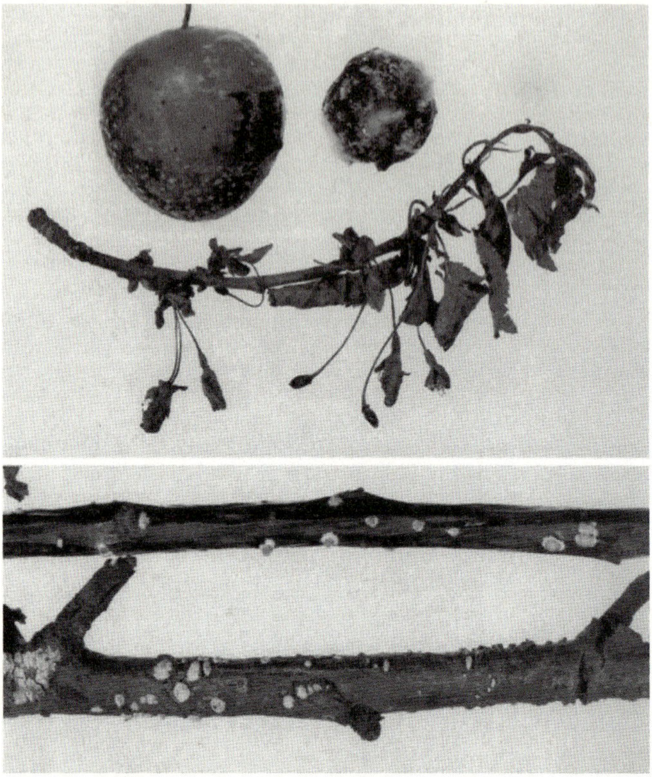

Plate 12 Brown Rot of Plums

Part II

Disease Control and Management

Control of a plant disease means reduction in the amount of damage caused. Our present annual toll from disease is nearly four billion dollars. Perfect control is rare, but profitable control, when the increased yield more than covers the cost of chemicals and labor, is quite possible. Commercial growers now average a return of four dollars for each dollar so invested. Keeping home plantings ornamental yields a large return in satisfaction and increased property value.

The five fundamental principles of control are exclusion, eradication, protection, resistance, and therapy.

1. *Exclusion* means preventing the entrance and establishment of pathogens in uninfested gardens, states, or countries. For home gardeners it means using certified seed or plants, sorting bulbs before planting, discarding any that are doubtful, possibly treating seeds or tubers or corms before they are planted, and, most especially, refusing obviously diseased specimens from nurseryman or dealer. For states and countries, exclusion means quarantines, prohibition by law. Sometimes restricted entry of nursery stock is allowed, the plants to be grown in isolation and inspected for 1 or 2 years before distribution is permitted.
2. *Eradication* means the elimination of a pathogen once it has become established on a plant or in a garden. It can be accomplished by *removal* of diseased specimens, or parts, as in roguing to control virus diseases or cutting off cankered tree limbs; by *cultivating* to keep down weed hosts and deep ploughing or spading to bury diseased plant debris; by *rotation* of susceptible with nonsusceptible crops to starve out the pathogen; and by *disinfection*, usually by chemicals, sometimes by heat treatment. Spraying or dusting foliage with sulfur after mildew mycelium is present is eradication, and so is treating the soil with chloropicrin to kill nematodes and fungi.
3. *Protection* is the interposition of some protective barrier between the susceptible part of the suscept or host and the pathogen. In most instances this is a protective spray or dust applied to the plant in advance of the arrival of the fungus spore; sometimes it means killing insects or other inoculating agents; sometimes it means the erection of a windbreak or other mechanical barrier.

R.K. Horst, *Field Manual of Diseases on Trees and Shrubs*,
DOI 10.1007/978-94-007-5980-0_2,

4. *Resistance* is control by the development of resistant varieties. Resistant varieties are as old as time. Nature has always eliminated the unfit, but since about 1,890 man has been speeding up the process by deliberately breeding, selecting, and propagating plants resistant to the more important diseases. Resistant ornamental plants have lagged behind food plants, but we do have wilt-resistant asters, rust-resistant snapdragons, wilt-resistant mimosas. Here is the ideal way for home gardeners to control their plant diseases – in the winter when the seed order and the nursery list is made out – so easy, and so safe!
5. *Therapy* is control by inoculating or treating the plant with something that will inactivate the pathogen. Chemotherapy is the use of chemicals to inactivate the pathogen, whereas heat is sometimes used to inactivate or inhibit virus development in infected plant tissues so that newly developing tissue may be obtained which is free of the pathogen.

A **fungicide** is a substance that destroys or inhibits the growth of fungi. It may be an **eradicant**, applied to a plant, plant part, or the environment as a curative treatment to destroy fungi established within a given area or plant; or preferably it may be a **protectant**, applied to protect a plant or plant part from infection by killing, or inhibiting the development of, fungal spores of mycelium that may arrive at the infection court. A **bactericide** is a substance that destroys or inhibits bacteria and nematicide for nematodes. Among the more recent bactericides are antibiotics, products of other living organisms. They also have value against certain fungi. There are few **virocides**, which are toxic or poisonous to viruses.

A **pesticide** is any chemical that is used to kill pests, especially insects and rodents.

An **insecticide** is an agent used to kill insects.

A **disinfectant** is an agent that frees a plant or plant part from infection by destroying the pathogen established within it. A disinfectant kills or inactivates organisms present on the surface of the plant or plant part or in the immediate environment. Chemicals for seed treatment can be either eradicants or protectants, but most of them are disinfectants, in that they kill organisms on the surface of the seed rather than those within. In common usage, however, they are called disinfectants.

A **nematicide** is, of course, a chemical that kills nematodes in the soil or in the plant. Most nematicides are **fumigants**, chemical toxicants that act in volatile form.

The search for new fungicides goes on, with hundreds of synthetic organic compounds being screened each year. This screening is often a cooperative venture between manufacturers, state experiment stations, and the U.S. Department of Agriculture. After safety precautions for the operator and the environment, and the effectiveness of a compound for certain diseases have been determined, the chief question is whether the material is **phytotoxic**, that is, injurious to plants, at concentrations required for control. Phytotoxicity is an elusive factor, not to be pinned down in a few tests. It varies not only with the kind of plant but with the particular variety, the amount of moisture in the soil when the spray is applied, the temperature, whether or not the application is followed by rain or high humidity, the section of the country, and the compatibility of the chemical with spreaders or wetting agents, as well as with other fungicides or insecticides. Coordinated tests with new materials in many different states are extremely valuable. Some compounds give rather uniform results over the country; others vary widely with climatic conditions.

Consumers are well protected against fraud, but they must be willing to read the fine print on labels if they are to choose intelligently from the bewildering array of proprietary compounds on dealers' shelves. Consumers must also read the fine print and follow directions exactly if their homegrown vegetables are to be as safe for consumption as those from commercial growers who have to comply with the law in the matter of residues.

Even if you follow exactly the directions for dosage given on the label, you may have some plant injury under your particular combination of soil, weather, and kinds of plants. Keep a notebook. Put down the date you sprayed, the dosage used, the approximate temperature and humidity, whether it was cloudy or sunny, in a period of drought or prolonged wet weather. Go around later and check for burning; for leaf spotting and defoliation from the spray or from failure to control the disease; for leaf curling or stunting; for too much unsightly residue. Note which varieties can take the spray and which cannot. Exclusive reliance must be placed on directions and information supplied by the manufacturer or by agricultural specialists, agents, or advisors. **Be sure to read the label**.

Part III

Host Plants and Their Diseases

Headings Under Which Diseases Are Described

ANTHRACNOSE – dead spots with definite margins, often with pinkish slimy spore masses, on leaves, stems, or fruit.
BACTERIAL DISEASES – all types of diseases, galls, blights, rots, leaf spots, caused by bacteria.
BLACK KNOT – black, knotty enlargement of woody tissue.
BLACKLEG – darkening at the base of a plant.
BLACK MILDEW – superficial dark growth caused by parasitic fungi.
BLACKSPOT – a dark leaf spot on rose.
BLIGHTS – general killing of leaves, flowers, stems.
BLOTCH DISEASES – irregular necrotic areas on leaves or fruit.
BROOMRAPES – leafless herbs parasitic on roots.
CANKERS AND DIEBACK – localized lesions on stems or trunks, sometimes accompanied by dying back from the top.
CLUB ROOT – distorted swollen roots.
DAMPING-OFF – sudden wilting of seedlings or rotting of seeds in soil.
DODDER – parasitic seed plant with orange tendrils.
DOWNY MILDEWS – with internal mycelium but fruiting structures protruding to form white, gray, or violet patches.
FAIRY RINGS – mushrooms growing in circles.
FRUIT SPOTS – blemishes on fruit.
GALLS – noticeable enlargements of leaves, stems, or roots.
LEAF BLISTER, LEAF CURL DISEASES – leaf deformities.
LEAF SCORCH – discoloration as if by intense heat.
LEAF SPOTS – delimited dead areas in leaves.
LICHENS – occasional on trees, or shrubs.
MISTLETOE – semiparasitic seed plant, forming leafy tufts in trees.
MOLDS – conspicuous fungus growth on leaves, seeds, or grafts.
NEEDLE CASTS – conspicuous shedding of evergreen foliage.
NEMATODES – causing decline diseases and flowers.

R.K. Horst, *Field Manual of Diseases on Trees and Shrubs*,
DOI 10.1007/978-94-007-5980-0_3, © Springer Science+Business Media Dordrecht 2013

NONPARASITIC DISEASES – due to environmental conditions rather than specific organisms.

POWDERY MILDEWS – superficial white felty or powdery growth on leaves and flowers.

ROTS – soft or hard decay or disintegration of plant tissues.

RUSTS – with reddish or rust-colored spore masses.

SCAB – raised or crustlike lesions on leaves or fruit.

SCURF – flaky or scaly lesions.

SLIME MOLDS – found in lawns.

SMUTS – with sooty black spore masses.

SNOWMOLD – light patches in turf, especially early spring.

SOOTY MOLD – superficial black mycelium growing in insect exudate.

SPOT ANTHRACNOSE – light spots with raised darker borders or scabby lesion caused by *Elsinoë* species.

VIRUS DISEASES – mosaics, ring spots, yellows, wilt caused by viruses.

WHITE RUSTS – white blisters in leaves.

WILTS – systemic diseases, with wilting, death of leaves, and branches.

WITCHWEED – weed parasitic on roots. The numbers given with the rusts refer to spore stages.

A

ABELIA

LEAF SPOT. **Cercospora abeliae**, IL.

NEMATODE, Root Knot. **Meloidogyne arenaria**; **M. hapla**; **M. incognita**.

NONPARASITIC. **Chlorosis**, due to soil alkalinity, TX.

POWDERY MILDEW. **Oidium** sp., TX.

ROT, Root. **Phymatotrichum omnivorum**, TX; **Rhizoctonia solani**, TX; **Pythium** sp.

VIRUS. **Abelia Latent**, MD.

ABUTILON (Flowering Maple, Indian Mallow, Velvet Leaf)

BLIGHT, Foliage. **Colletotrichum coccodes**, VT.

BLIGHT, Phytophthora. **Phytophthora capsici**, IL.

LEAF SPOT. **Alternaria** sp., IL, IN, NJ, NY, TX; **Cercospora avicennae**, MO, VA; **Cladosporium herbarum**, KS, NY; **Colletotrichum malvarum**, IA; **Phyllosticta althaeina**, TX.

NEMATODE, Root Knot. **Meloidogyne** sp., AL, FL; **M. incognita**, IN; **M. hapla**.

ROT, Root. **Phymatotrichum omnivorum**, TX; **Armillaria mellea**, CA.

ROT, Stem. **Macrophomina phaseoli**, IL.

RUST. **Puccinia heterospora** (III), FL to AZ.

VIRUS. **Abutilon Infectious Variegation**; **Abutilon Yellows**, CA; **Bean Mosaic**, HI; **Mosaic**, universal.

WILT. **Verticillium albo-atrum**, NJ; **V. dahliae**, WI.

The leaf spots and rot occur on Indian mallow and velvet leaf. Variegation in flowering maple is a true mosaic disease, although fostered as a desirable ornamental quality.

ACACIA

CANKER, Twig and Branch. **Nectria ditissima**, CA, SC.

LEAF SPOT. **Physalospora fusca**, FL; **Cercospora** sp., GA; **Phyllachora texana**, TX.

LEAF SPOT, Algal. **Cephaleuros virescens**, FL.

R.K. Horst, *Field Manual of Diseases on Trees and Shrubs*,
DOI 10.1007/978-94-007-5980-0_4, © Springer Science+Business Media Dordrecht 2013

MISTLETOE. **Phoradendron californicum**, CA, TX; **P. serotinum** (**flavescens**), TX.
NEMATODE, Root Knot. **Meloidogyne** sp.
NONPARASITIC. **Chlorosis**, due to excess lime, CA.
Gummosis, due to deficient or irregular moisture, CA.
POWDERY MILDEW. **Erysiphe polygoni**, CA.
ROT, Heartwood. **Ganoderma applanatum**, CA.
ROT, Mushroom Root. **Armillaria mellea**, occasional; **Clitocybe tabescens**, FL.
ROT, Root. **Phymatotrichum omnivorum**, TX.
ROT, Sapwood. **Schizophyllum commune**, CA.
RUST. **Ravenelia australis**, TX; **R. gooddingii**, AZ; **R. hieronymi**, witches' broom, TX; **R. igualica**, TX; **R. roemerianae**, TX; **R. siliquae**, NM, TX; **R. subtortuosae**, witches' broom, TX; **R. thornberiana**, witches' broom, AZ, TX; **R. versatilis**, AZ, CA, NM, TX.

ACACIA (Koa Tree)
LEAF SPOT, Brown; Dark Blight. **Calonectria colhounii**, HI.

ACALYPHA (Copper-Leaf)
DOWNY MILDEW. **Plasmopara acalyphae**, WI.
GALL, Leaf. **Synchytrium aureum**, WI.
LEAF SPOT. **Cercospora acalyphae**, NY to AL, OK, TX, WI; **Phyllosticta** sp., NJ; **Ramularia acalyphae**, TX.
NEMATODE, Lesion. **Pratylenchus** sp., FL.
NEMATODE, Root Knot. **Meloidogyne** sp., GA, MD; **M. incognita**, IN.
POWDERY MILDEW. **Golovinomyces cichoracearum**, WI.
ROT, Mushroom Root. **Clitocybe tabescens**, FL.
ROT, Root. **Phymatotrichum omnivorum**, TX; **Rhizoctonia solani**, IL.

ACANTHOPANAX (Five-Leaf Aralia)
LEAF SPOT. **Alternaria** sp., MO.
ROT, Root. **Phymatotrichum omnivorum**, TX.

AESCHYNANTHUS
VIRUS. **Tobacco Mosaic**, CA, CT, DC, FL, OH, WA.

AILANTHUS (Tree-of-Heaven)
BLACK MILDEW. **Dimerosporium robiniae**, DC.
BLIGHT, Twig. **Gibberella baccata** (*Fusarium lateritium*), VA; **Diplodia ailanthi**, TX; **D. natalensis**, TX; **Phoma ailanthi**, TX.
CANKER; DIEBACK. **Nectria cinnabarina**, KS, NJ, SC; **N. coccinea**, occasional; **Physalospora obtusa**, KS, MI, NY.
LEAF SPOT. **Cercospora glandulosa**, widespread; **Gloeosporium ailanthi**, LA, TX, WV; **Phyllosticta ailanthi**, VA; **Cristulariella pyramidalis**, FL.
ROT, Butt. **Daedalea unicolor**, occasional.
ROT, Root. **Armillaria mellea**, Northeast; **Phymatotrichum omnivorum**, TX.
ROT, Wound. **Polyporus lacteus**; **P. versicolor**, occasional in living trees; **Schizophyllum commune**.

Wilt. **Verticillium albo-atrum**, NY, PA, VA.

Ailanthus is well adapted to city smoke and not often troubled by disease except for wilt, which can be serious and has killed many trees in the Philadelphia area.

AKEE TREE (*Blighia sapida*)
Wilt; Dieback. **Verticillium dahliae**, FL.

ALDER (*Alnus*)
Bacterial Leaf Blight. **Xanthomonas campestris**, CA.
Blight, Gray Mold. **Botrytis cinerea**, WA.
Canker, Brown Felt. **Septobasidium filiforme**, NC; **S. peckii**, NY.
Canker; Dieback. **Nectria coccinea**, NY, PA; **N. galligena**, NH; **Hymenochaete agglutinans**, MI, NY, PA; **Melanconis alni**, ME to NJ, WI; **Phomopsis alnea**, KY; **Physalospora obtusa**, SC, VA; **Solenia ochracea**, widespread on bark; **Didymosphaeria oregonensis**, ID, OR, WA; **Valsa melanodiscus**, western North America.
Canker, Stem. **Cytospora** sp., OR.
Dodder. **Cuscuta compacta**, MD, VA.
Gall, Root. Cause unknown.
Leaf Blister, Catkin Hypertrophy. **Taphrina robinsoniana**, widespread; **T. japonica**, AK, CA, OR; **T. occidentalis**, CA, ID, MT, OR, WA.
Leaf Spot. **Cercospora alni**, WI; **Cercosporella alni**, AK, OR, WA; **Dothidella alni**, black spot; **Gloeosporium tubiformis**; **Cylindrosporium vermiforme**, WI; **Ophiodothis alnea**, tar spot, NH, PA; **Phyllosticta alnea**, AK; **Septogloeum variegatum**, CA; **Septoria alni**, AK, CA, NY, WA, WI, OR; **Hypospila californica**, CA.
Mistletoe. **Phoradendron serotinum** (**flavescens**), AZ, NM; **P. villosum**, CA; **Viscum album**, CA; on *Alnus rubra* (Red Alder).
Mistletoe, European. **Viscum album**, CA.
Nematode, Root Knot. **Meloidogyne** sp.
Powdery Mildew. **Erysiphe aggregata**, ME to NJ, CA, MN, WA; **Phyllactinia corylea**, ME to WI, AL, SC; **Microsphaera alni**, widespread.
Rot, Heart. **Daedalea unicolor**; **Fomes igniarius**; **F. pinicola**; **Ganoderma applanatum**; **Polyporus sulphureus**.
Rot, Root. **Armillaria mellea**, CA; **A. nabsnona**, HI; **Phymatotrichum omnivorum**, TX.
Rot, Wood. **Daedalea confragosa**; **Daldinia occidentalis**; **Fomes scutellatus**; **Lenzites saepiaria**; **L. trabea**; **Pholiota adiposa**; **Pleurotus serotinus**; **Polyporus adustus**; **P. hirsutus**; **P. versicolor**; **Schizophyllum commune**; **Steccherinum ochraceum**; **Stereum** spp., mostly on dead wood.
Rust. **Melamsoridium alni** (II, III), CA, TX.
Sooty Mold. **Scorias spongiosa**, NY to NC, WV.

ALLSPICE (*Pimenta*)
Rust. **Puccinia psidii**, FL.

ALMOND (*Prunus amygdalus*)
Anthracnose. **Colletotrichum acutatum**, CA.

ANTHRACNOSE; Kernel Rot. **Gloeosporium amygdalinum**, CA.
BACTERIAL Crown Gall. **Agrobacterium tumefaciens**, AL, AZ, CA, NC.
BACTERIAL Leaf Spot. **Xanthomonas pruni**, NE.
BACTERIAL Shoot Blight; Blast. **Pseudomonas syringae**, CA.
BLIGHT, Blossom; Brown Rot. **Monilinia laxa**, CA, OR; **Botrytis cinerea**.
BLIGHT, Leaf; Shot-Hole Disease. **Coryneum carpophilum**, CA, ID, OR.
CANKER; DIEBACK; Crown Rot. **Phytophthora cactorum**, CA, **P. citrophthora**, CA; **Botryosphaeria dothidea** (Dothiorella canker) CA; **Ceratocystis fimbriata**, CA.
CANKER, Pruning Wound. **Phytophthora syringae**, CA.
DECLINE, Almond. Unknown etiology, CA. “Golden Death”.
LEAF SPOT. **Cercospora circumscissa**, CA, OR; **Hendersonia rubi**, CA; **Alternaria alternata**, CA; **A. arborescens**, CA; **A. tenuissima**, CA.
NEMATODE, Lesion. **Pratylenchus vulnus**, CA.
NEMATODE, Root Knot. **Meloidogyne** sp., AZ.
NONPARASITIC. **Bud Failure**. Seed transmitted, increased by pruning.
Chlorosis. Iron deficiency.
Little Leaf. Zinc deficiency.
POWDERY MILDEW. **Podosphaera tridactyla**, CA; **Sphaerotheca pannosa**, CA.
ROT, Fruit. **Phomopsis amygdali**, CA, and associated branch dieback.
ROT, Green Fruit. **Sclerotinia sclerotiorum**, CA.
ROT, Heart. **Polyporus versicolor**, OR.
ROT, Hull. **Rhizopus** spp.
ROT, Root. **Armillaria mellea**, CA, NC; **Phymatotrichum omnivorum**, TX.
ROT, Wound. **Schizophyllum commune**, CA.
RUST. **Tranzschelia discolor** (II, III); 0, I on anemone, CA.
SCAB. **Cladosporium carpophilum**, CA, CT, OR.
VIRUS. **Almond Bud Failure**; **Almond Calico**, CA; **Peach Mosaic**; **Peach Ring Spot**; **Peach Rosette**; **Peach Yellow Bud Mosaic**; **Peach Yellows**.
WILT. **Verticillium albo-atrum**, CA, IL.

Crown rot and Armillaria rot are often limiting factors in almond production. At least five sprays are necessary to control brown rot, leaf blights, and scab, and often a zinc sulfate spray for little leaf. The California Agricultural Experiment Station Extension Service provides, each year, a revised “Spray, Dust and Fumigation Program for Almonds”.

ALMOND, FLOWERING (*Prunus triloba*)
BACTERIAL Fire Blight. **Erwinia amylovora**, IN.
BACTERIAL Spot. **Xanthomonas pruni**, NJ.
BLIGHT, Blossom and Twig. **Botrytis cinerea**, NY; **Monilinia fructicola**, CT, KS.
POWDERY MILDEW. **Podosphaera oxyacanthae**, IA.
ROT, Mushroom Root. **Armillaria mellea**, MS; White Root, **Corticium galactinum**, MD.

AMELANCHIER (Serviceberry, Juneberry)
BACTERIAL Fire Blight. **Erwinia amylovora**, widespread.

Black Mildew; Witches Broom. **Apiosporina collinsii**, widespread.
Blight, Leaf. **Fabraea maculata**, widespread.
Blight, Fruit and Leaf. **Monilinia gregaria**, CO, IA, WA.
Canker. **Nectria cinnabarina**, occasional; **Pezicula pruinosa**, wide-spread; **Gloeosporium perennans**, OR.
Canker, Blister. **Nummularia discreta**, IA, KS, MA, OH.
Leaf Blister; Witches' Broom. **Taphrina amelanchieri**, CA.
Leaf Spot. **Coccomyces tumidus**, MO, MT; **Cristulariella pyramidalis**, NC; **Phyllosticta innumerabilis**, IA, MN, NE, ND; **P. paupercula**, KS, MT; **P. virginiana**, NY, PA, WI.
Powdery Mildew. **Erysiphe polygoni**, ID, MT; **Phyllactinia corylea**, OR, WA; **Podosphaera oxyacanthae**, occasional.
Rot, Brown; Fruit. **Monilinia amelanchieris**, NY; **M. fructicola**, MI, OH.
Rot, Root. **Phymatotrichum omnivorum**, TX.
Rot, Wood Butt. **Daedalea unicolor**, cosmopolitan.
Rust. **Gymnosporangium biseptatum** (I); III on *Chamaecyparis*, northeastern and Middle Atlantic States. Horned galls, underside leaves. **G. clavariiforme** (I) on leaves, fruits, stems; III on juniper. **G. clavipes** (I) on fruits, stems; III on juniper. **G. corniculans** (I); III on juniper. **G. cupressi** (0, I); III on cypress. **G. gracile** (0, I); III on juniper. **G. harknessianum** (I) on fruits, CA, OR; III on juniper. **G. inconspicuum** (I) on fruits; III on juniper. **G. juvenescens** (I); III on juniper, in West (may be **G. nidusavis**). **G. kernianum** (I) on leaves, AZ, CO, OR; III on juniper. **G. libocedri** (I) on leaves, CA, OR; III on juniper. **G. nelsonii** (I) on leaves, fruits; III on juniper, MI. **G. nidusavis** (I) on leaves, fruits, stems; III on juniper, central and eastern states.

AMORPHA (Leadplant, Indigobush)
Canker, Twig. **Cytospora amorphae**.
Leaf Spot. **Cylindrosporium passaloroides**, widespread; **Diplodia amorphae**, TX.
Powdery Mildew. **Erysiphe polygoni**, IA, MN, WY.
Rust. **Uropyxis amorphae**, general.

AMPELOPSIS (*A. cordata* and other species)
Canker; Dieback. **Nectria cinnabarina**, occasional.
Downy Mildew. **Plasmopara viticola**, NY, WI.
Leaf Spot. **Cercospora truncata**, LA; **C. vitis**, AL, LA; **Guignardia bidwellii**, widespread; **Linospora psederae**, WV; **Phloeospora ampelopsidis**, WI.
Nematode, Dagger. **Xiphinema index**.
Powdery Mildew. **Uncinula necator**, widespread.
Rot, Root. **Helicobasidium purpureum**, TX.

ANDROMEDA (Bog-Rosemary)
(See Lyonia and Pieris for shrubs commonly called Andromeda)
Black Mildew. **Asterina clavuligera**, GA, NJ.
Blight, Ramorum. **Phytophthora ramorum**, PA.
Gall; Red Leaf Spot. **Exobasidium vaccinii**, AK, ME, NY, WI.

LEAF SPOT. **Venturia arctostaphyli**, NY.
LEAF SPOT. Tar. **Rhytisma andromedae**, ME to NC, AK, MI, MN.
POWDERY MILDEW. **Microsphaera alni** var. **vaccinii**, MA, NJ.
RUST. **Chrysomyxa cassandrae** (II, III); 0, I on pine.

ANGELONIA (Angelonia angustifolia)
NEMATODE, Root Gall, **Meloidogyne mayaguensis**, FL.
VIRUS, **Alternanthera Mosaic**, NY.

ANISE-TREE (*Illicium floridanum*)
BLACK MILDEW. **Lembosia illiciicola**, AL, MS.
LEAF SPOT, Algal; Green Scurf. **Cephaleuros virescens**, LA, SC.
SOOTY MOLD. **Capnodium footii**, MS.

ANTIDESMA (Chinese-Laurel)
LEAF SPOT, Algal; Green Scurf. **Cephaleuros virescens**, FL.

APPLE (*Malus sylvestris*)
ANTHRACNOSE, Northwestern; canker; Fruit Rot. **Neofabraea malicorticis**, prevalent OR, WA, occasional CA, IL, MA, ME, MI, NE, OK.
BACTERIAL Blast, of flowers, shoots. **Pseudomonas syringae**, AR, CA; Blister Spot of bark and fruit. **P. syringae** pv. **papulans**; Rot, **P. melophthora**.
BACTERIAL Crown Gall. **Agrobacterium tumefaciens**, general; Hairy Root, **A. rhizogenes**, Central States, ID, NY.
BACTERIAL Fire Blight. **Erwinia amylovora**, general.
BLIGHT, Limb. **Corticium laetum**, LA, NC; **C. salmonicolor**, FL, LA.
BLIGHT, Southern. **Sclerotium rolfsii**, on seedlings.
BLIGHT, Thread. **Pellicularia koleroga**, WV and IN to Gulf states.
BLOTCH, Fruit; LEAF SPOT; Twig Canker. **Phyllosticta solitaria**, general except New England and the far South.
BLOTCH, Sooty, of fruit. **Gloeodes pomigena**, eastern and central states; **Geastrumia polystigmatis**, NC.
CANKER, Bark; Fruit Rot. **Myxosporium corticola**, New England to MD, IL, MI, OK, OR, SD.
CANKER, Blister. **Nummularia discreta**, east of the Rocky Mts.
CANKER, Bole. **Cylindrocarpon didymium**, WA; **Leucostoma cincta**. WI.
CANKER, Crater. Cause unknown, sunken bark, crown rot, decline. WA.
CANKER; DIEBACK. **Botryosphaeria ribis** (syn. **B. dothidea)**, East and South; **Cytospora** spp., widespread; **Fusarium** spp., also Fruit Rot, Pacific Northwest, Secondary to drought, winter injury, insect punctures; **Coryneum foliicola**, widespread; **Glutinium macrosporium**, OR; **Leptosphaeria coniothyrium**, East, central states, WI; **Plenodomus fuscomaculans**, CA, MI; **Sphaeropsis pyriputrescens**, WA.
CANKER, European. **Nectria galligena**, East and central states to NC and MS; Pacific Coast.
CANKER; Leaf Spot. **Monochaetia mali**, IL, NJ to SC, MO, WV; **Leucostoma cincta**, MI.

CANKER, Perennial; Bull's-Eye Fruit Rot. **Neofabraea perennans**, ID, MT, OR, WA.

CANKER; Sapwood Rot. **Hymenochaete agglutinans**, CT; **Hypoxylon** spp.

CANKER, Scurf Bark. **Clasterosporium** sp.

CANKER, Twig. **Coryneum foliicola**, widespread; **Gibberella baccata**; **Nectria cinnabarina**, northern U.S.

DECLINE. **Irpex tulipiferae**, MN; **Coriolus versicolor**, also dieback, MN, WA; **Schizophyllum commune**, MN.

FRUIT SPOT, Black Pox; Canker; Leaf Spot. **Helminthosporium papulosum**, GA, IN, MA, MS, NJ, OH, PA, WV; **Phacidiopycnis washingtonensis** (also Fruit Rot), WA.

FRUIT SPOT, Brooks. **Mycosphaerella pomi**, New England to NC, AR, IA, MO.

FRUIT SPOT, Flyspeck. **Microthyriella rubi**, general.

LEAF SPOT. **Cercospora mali**, Gulf states, VT; **Coniothyrium pyrinum**; **Diaporthe pernisiosa**, AR; **Fabraea maculata**, IA; **Illosporium malifoliorum**, PA to NC, IN; **Mycosphaerella sentina**, IL, NJ, PA; **Pestalotia** spp., MD to NC, ID, IN; Ghost Spot, **Alternaria tenuis**, GA; **Cristulariella pyramidalis**, FL, on *Malus pumila*, **Glomerella cingulata**, TN.

MISTLETOE. **Phoradendron serotinum (flavescens)**, NC, TX; **Viscumalbum**, CA.

NEMATODE, Lesion. **Pratylenchus penetrans**; **P. pratensis**.

NEMATODE, Root Knot. **Meloidogyne** spp., MS, TX, UT.

NONPARASITIC. **Bitter Pit**, **Stippen**, **Baldwin Spot**, general in storage.

Black End. Probably drought injury, AL, OR, WA.

Black Heart, of wood. Perhaps freezing injury; widespread.

Brown Core, in MacIntosh variety. Excessive nitrogen, low storage temperature.

Callus Knot. Wound overgrowth from defective union of stock and scion.

Chlorosis. Usually iron deficiency in alkaline western soils.

Collar Rot. Winter injury.

Cork; **Rosette**. Boron deficiency, northeastern states to IN, KY, Pacific Coast.

Internal Bark Necrosis. Cause unknown, possibly same as measles; general.

Jonathan Spot. Associated with dry weather, delayed storage.

Leaf Scorch. Magnesium deficiency.

Measles, reddish pimples in bark. Widespread; cause unknown, possibly boron deficiency in part. False measles due to manganese toxicity.

Ozone Injury. Pitted area in flesh.

Rosette; Little Leaf. Usually zinc deficiency.

Scald. Discoloration of fruit skin by volatile respiratory products, general in storage. Controlled by oiled paper wraps.

Soft Scald and **Soggy Breakdown**. Associated with delayed storage and low temperature.

Stigmonose. Fruit dimpling, distortion from insect punctures; widespread.

Sunburn, of fruit. Heat or light injury.

Sunscald, of bark. Freezing injury of trunk and larger branches on side exposed to sun.

Water-Core. Deficient or irregular water supply.

POWDERY MILDEW. **Podosphaera leucotricha** and **P. oxyacanthae**, general.
ROT, Bitter, of Fruit; Canker. **Glomerella cingulata**, general, especially in South, rare in West; **G. rubicola**, IL; **Colletotrichum acutatum**, AR, MI.
ROT, Black; Frog-Eye Leaf Spot; Canker. **Physalospora obtusa**, general to the Great Plains; **P. mutila**, CA, MT, OR, WA; **P. rhodina**, KY.
ROT, Black Root. **Botryosphaeria obtusa**, PA; **Xylaria mali**, eastern and central states.
ROT, Brown. **Monilinia fructicola**, general except for far South; **M. laxa**, OR, WA.
ROT, Calyx-end. **Alternaria** sp., **Sclerotinia sclerotiorum**, NH, WA.
ROT, Crown. **Phytophthora cactorum**, MA.
ROT, Fruit. **Alternaria mali**; **Aspergillus niger**; **Cephalosporium carpogenum**; **Corticium centrifugum**, Pacific Northwest, occasional in East; **Chaetomella** sp., WA; **Cladosporium** spp.; **Endomyces mali**; **Epicoccum granulatum**, WA; **Gloeosporium** spp.; **Gliocladium viride**, IL; **Hormodendron cladosporiodes**, WA; **Mucor piriformis**, CA, WA; **Oospora** sp.; **Phialophora malorum**, IN, VA, WA; **Pleospora fructicola**, WA; **P. herbarum**, Pacific states; **Penicillium** spp., cosmopolitan; **Phoma** spp.; **Stemphylium congestum**, WA; **Tricothecium roseum**, general; **Botrytis** spp.; **Colletotrichum fructus**; **Rhizopus nigricans**, cosmopolitan; **Phytophthora syringae**, OR.
ROT, Fruit (Postharvest). **Botrytis cinerea**, **Phomopsis mali**, core rot, CA, NY; **Trichoderma harzianum**, MD; **Sphaeropsis pyriputrescens**, WA; **Phacidiopycnis washingtonensis**; **Potebniamyces pyri**, WA.
ROT, Heartwood. **Fomes fomentarius**; **F. igniarius**; **F. pinicola**; **Ganoderma applanatum**.
ROT, Mushroom Root. **Armillaria mellea**, prevalent on Pacific Coast; **Clitocybe tabescens**, AR, FL, TX, VA.
ROT, Postharvest Speck. **Phacidiopycnis washingtonensis**, WA.
ROT, Root. **Botryodiplodia theobromae**, AL; **Hymatotrichum omnivorum**, AK, AZ, NM, TX; **Fusarium oxysporum**, ID; **F. solani**, ID.
ROT, Silverleaf. **Stereum purpureum**, KS, ME, MN, NY, Pacific Northwest.
ROT, White Root. **Corticium galactinum**, DE to VA; AR, IL; **Rosellinia necatrix**, CA.
ROT, Wood. **Daedalea confragosa**; **Poria** spp.; **Pleurotus ulmarius**; **Pholiota adiposa**; **Polyporus** spp.; **Stereum** sp.; **Schizophyllum commune**; **Trametes** spp.
RUSSET, Fruit. **Aureobasidium pullulans** and **Rhodotorula glutinis**, NY.
RUST. **Gymnosporangium libocedri**. (0, I), on leaves, fruit; III on incense cedar, CA, OR; **Gymnosporangium nidus-avis** (0, I), on leaves, fruit, stems, IN, MD, MS, NJ; III on red-cedar.
RUST, Cedar-Apple. **Gymnosporangium juniperivirginianae** (0, I) on leaves, fruit; general east of Great Plains; III on red-cedar and Rocky Mountain juniper.
RUST, Hawthorn. **Gymnosporangium globosum** (0, I) on leaves; III on red-cedar, ME to AK, KS, NE.

RUST, Quince. **Gymnosporangium clavipes** (0, I), on fruit; III on common juniper and red-cedar.
SCAB. **Venturia inaequalis**, general.
SOOTY MOLD. **Fumago vagans**, occasional.
SPOT ANTHRACNOSE. **Elsinoë piri**, OR, WA.
VIRUS. **Apple Chlorotic Leaf Spot**; **Apple Stem Growing**; **Apple Mosaic**; **Tulare Apple Mosaic**, CA; **Tobacco Mosaic**; **Dapple Apple**, NH; **Green Mottle**, NY; **Stem-pitting**; **Tomato Ringspot**, NY; **Decline** and **Graft Union Necrosis**, WA.
WILT. **Verticillium albo-atrum**.

If this appalling list of diseases should make you think twice before planting apples in the backyard with the expectation of getting cheap and abundant fruit, that is all to the good. There is no easy, or cheap, road to perfect fruit. The commercial grower may, in a wet season, apply nearly 20 sprays to keep scab under control. The homeowner thinks one or two are enough. State experiment stations offer abbreviated schedules for home gardeners, tailored for the area, and if these are followed carefully, a fairly good crop can be expected. There are also all-purpose mixtures available for fruit trees, which may work reasonably well. Scab is the most important apple disease, and proper timing of early season sprays is most essential. Bitter rot, black rot, sometimes rust need attention. Fire blight control is primarily a question of proper pruning to remove infected wood, with an antibiotic spray during bloom.

APRICOT (*Prunus armeniaca*)
BACTERIAL Canker; Gummosis. **Pseudomonas syringae**, CA, OR; **Criconema mutabile**, CA.
BACTERIAL Crown Gall. **Agrobacterium tumefaciens**, widespread. Japanese apricot (*Prunus mume*) is resistant.
BACTERIAL Fire Blight. **Erwinia amylovora**, CO, FL, NE, PA, TX.
BACTERIAL Leaf Spot. **Xanthomonas pruni**, IL to TX, NE.
BLACK KNOT. **Dibotryon morbosum**, CO, IA, NY.
BLIGHT, Blossom, Twig; Brown Rot. **Monilinia laxa**, Pacific Coast states.
BLIGHT, Shoot; Shot Hole; Fruit Spot. **Coryneum carpophilum**, wide-spread; **Monilinia laxa**, MA, NY, RI.
CANKER, Felt Fungus. **Septobasidium pseudopedicellatum**, MS.
CANKER, Trunk. **Phytophthora cactorum**, CA; **P. citrophthora**, CA.
CANKER, Trunk and Limb Gall. **Monochaetia rosenwaldia**, CA.
CANKER, Twig; DIEBACK. **Cytospora** spp., AZ, NY; **Valsa leucostoma**, MO, TX; Cytosporina Dieback, **Eutypa armeniaceae** (syn. **E. lata**), CA; **Ceratocystis fimbriata**, CA.
CANKER; DIEBACK; Coral Spot. **Nectria cinnabarina**, IN, WA.
FRUIT SPOT. **Venturia cerasi** (See under SCAB).
LEAF CURL. **Taphrina deformans**, SC.
LEAF SPOT. **Cercospora circumscissa**, TX; **Coccomyces** sp., CA, MA, TX, VT; **Phyllosticta circumscissa**, CA.
NEMATODE, Dagger. **Xiphinema** sp., WV, NY.
NEMATODE, Lesion. **Pratylenchus vulnus**, CA; **P. penetrans**, NY.

NEMATODE, Root Knot. **Meloidogyne** sp., AZ, TX.
NONPARASITIC. **Arsenical Injury**, from accumulation in soil, CA.
Chlorosis, alkali injury. Mineral deficiency, AZ, CA.
Exanthema. Copper deficiency, CA.
Gummosis, Sour Sap. Adverse soil and moisture conditions, AZ, CA, NJ, WA.
Internal Browning, Fruit Cracking. Boron deficiency, WA.
Little Leaf. Zinc deficiency, CA, OR, WA.
Mottle Leaf. Manganese deficiency, CA.
POWDERY MILDEW. **Podosphaera oxyacanthae**, CA, IA; **Sphaerotheca pannosa**, CA, NY.
ROT, Blossom End; Fruit Spot. **Alternaria** sp., and **A. citri** CA.
ROT, Brown; Blossom Blight. **Monilinia fructicola**, widespread on ripe fruit.
ROT, Green Fruit; Gummosis. **Botrytis cinerea**, CA.
ROT, Green Fruit; Twig Blight. **Sclerotinia sclerotiorum**, CA; **Lambertella pruni**, CA.
ROT, Heart. **Schizophyllum commune**, after freezing, TX, WA; **Trametes hispida**, CO.
ROT, Mushroom Root. **Armillaria mellea**, CA, TX; **Clitocybe tabescens**, FL.
ROT, Root. **Phymatotrichum omnivorum**, AZ, OK, TX.
ROT, Silver Leaf. **Stereum purpureum**, CA.
RUST. **Tranzschelia discolor** and **T. pruni** (II, III), CA, MS, NM, TX.
SCAB; Freckle; Twig Canker. **Cladosporium carpophilum**, wide spread.
VIRUS. **Apricot Gummosis**, WA; **Apricot Pucker Leaf**, UT; **Apricot Ring Pox**, CA, CO, WA; **Peach Mosaic**, Southwest; **Peach Phony**, GA; **Peach Ring Spot**, western U.S.; **Peach Rosette**; **Peach Yellow Bud Mosaic**, CA; **Peach Yellows**, occasional in East; **Stem Pitting**.
WILT. **Verticillium albo-atrum**, CA, UT, WA.

Apricots are very susceptible to Armillaria root rot and should be on resistant Myrobalan rootstock. Bacterial canker is an epidemic disease in many seasons, with activity starting in late autumn, ceasing in early summer. Sprays are usually needed in California for zinc deficiency, brown rot, green or jacket rot, and shot hole.

ARALIA, HERCULES CLUB (*Aralia spinosa*)
CANKER; DIEBACK. **Botryosphaeria ribis**, GA, VA, WV; **Nectria cinnabarina**, VA, WV.
LEAF SPOT. **Cercospora atromaculans**, LA, TX; **Phyllosticta araliae**, TX; **P. everhartii**, TX, WV; **Stagonospora** sp., WV.
ROT, Root. **Phymatotrichum omnivorum**, TX; **Rhizoctonia solani**, TX.
ROT, Sapwood. **Polyporus tulipiferae**, MD.
SPOT ANTHRACNOSE; Scab. **Sphaceloma araliae**, MD, MO.

ARALIA, MING (*Polyscias*)
BLIGHT. **Alternaria panax**, FL.

ARALIA, SARSAPARILLA, AMERICAN SPIKENARD (*Aralia hispida, A. nudicaulis, A. racemosa*)

LEAF SPOT. **Alternaria** sp., NY; **Ascochyta marginata**, WI; **Cercospora leptosperma**, IA, MI, NY, WI; **Phyllosticta decidua**, WI; **Ramularia repens**, WI; **Sclerotium deciduum**, WI.

POWDERY MILDEW. **Phyllactinia corylea**, MI, NE.

RUST. **Nyssopsora clavellosa** (III), CA, MN, OR, TX, ME to PA.

WILT. **Verticillium albo-atrum**, NY.

ARALIA, UDO (*Aralia cordata*)

BLIGHT. **Alternaria** sp., DE, NJ.

ROT, Stem. **Sclerotinia sclerotiorum**, MD.

WILT. **Verticillium albo-atrum**, MD, PA.

ARAUCARIA (Monkey-Puzzle, Norfolk-Island-Pine)

BACTERIAL Crown Gall. **Agrobacterium tumefaciens**, CA.

BLIGHT. **Cryptospora longispora**.

LEAF SPOT. **Pestalotia funerea**, CA, and **P. micheneri**, PA, probably secondary; **Stictis araucariae**, CA.

ROOT DISEASE. **Armillaria tabescens**, Mexico.

ROT, Root. **Cylindrocladium clavatum**, FL.

ARBORVITAE (*Thuja*)

BLIGHT, Fire. **Cercospora thujina**, on oriental arborvitae, AR, GA, LA, MS, TX.

BLIGHT, Gray Mold. **Botrytis cinerea**, NJ.

BLIGHT, Leaf. **Didymascella thujina**, TX, VT to WI.

BLIGHT, Nursery. **Phomopsis juniperovora**, IN, KY, OH, PA, VA; **P. occulta**, secondary.

BLIGHT, Snow. **Phacidium infestans**, Northeast.

BLIGHT, Twig. **Alternaria** sp.; **Mycosphaerella** sp.; **Pestalotia funerea**; **Phytophthora** sp.; **Heterosporium** sp., OR.

BLIGHT, Twig. **Coryneum berckmansii** on oriental arborvitae, OR, WA.

CANKER, Bark Patch. **Aleurodiscus amorphus**; **A. nivosus**, other species.

CANKER, Cypress. **Coryneum cardinale**, CA.

CANKER; DIEBACK. **Diplodia** sp., AL, FL.

DAMPING-OFF. **Rhizoctonia solani**, NM, NY, TX, VA.

NEEDLE CAST. **Lophodermium thujae**, ME, NH, NY, WI.

ROT, Mushroom Root. **Armillaria mellea**, MI, MS, NY, TX; **Clitocybe tabescens**, FL, LA.

ROT, Root. **Phymatotrichum omnivorum**, TX; **Fusarium solani**, TX.

ROT, Wood. **Fomes annosus**, MI; **F. roseus**, ME; **Poria vaporaria**, **P. weirii**, Great Lakes states; **Lenzites saepiaria**, MN; **Polyporus** spp.; **Schizophyllum commune**, ME; **Trametes** spp.

Giant arborvitae used for timber has many more wood rots than those listed. Oriental arborvitae in the South often looks as if blighted by fire, with nursery losses higher than in gardens. A copper spray, monthly from June to September, controls this blight. In the Northwest **Coryneum berckmansii** causes discoloration and shedding of branches, while **C. cardinale** is sporadically injurious.

ARDISIA

LEAF SPOT, Algal. **Cephaleuros virescens**, FL.

ARTEMISIA (Wormwood)

BLIGHT, Leaf. **Cercospora olivacea**, NJ, NY, ND; **Systremma artemisiae**, black pustule, PA.

DOWNY MILDEW. **Peronospora leptosperma**, IA, MN, ND, SD, WI.

GALL, Leaf. **Synchytrium aureum**, WI.

LEAF SPOT. **Cercospora ferruginae**, NY, WI; **Gloeosporium heterophyllum**, CA.

LEAF-HAIRS, DISCOLORATION. **Nematostoma artemisiae**, MO.

NEMATODE, Root Knot. **Meloidogyne** sp., FL.

POWDERY MILDEW. **Golovinomyces cichoracearum**, CA, IA, SD, TX.

ROT, Root. **Phymatotrichum omnivorum**, TX.

ROT, Root and Crown. **Fusarium solani**, CA.

RUST. **Puccinia tanaceti** (*P. absinthii*) (0, I, II, III), CA; **P. atrofusca** (0, I) ND; II, III on *Carex* spp.; **P. millefolii** (III), CA; **P. similis**, ID, OR.

VIRUS. **Tomato Spotted Wilt**, PA.

WHITE RUST. **Albugo tragopogonis**, IA, MT, ND, SD, TX, WI.

ARTOCARPUS (Breadfruit)

RUST. **Uredo artocarpi** (11), HI.

ASH (*Fraxinus*)

ANTHRACNOSE, Leaf Scorch. **Discula fraxinea**, DE, LA, OR; **Gloeosporium aridum**, eastern and central states.

BACTERIAL Decline. **MLO**, IL, IN, IA, NY.

BACTERIAL Hairy Root. **Agrobacterium rhizogenes**, IA, NE to OK.

BACTERIAL Phytoplasma. **Ash Yellows**, **Witches' Broom**, MI, MT, NE, NC, SD.

BLACK MILDEW. **Dimerosporium pulchrum** (*Sarcinella heterospora*).

BLIGHT, Seedling. **Rhizoctonia solani**, OK.

CANKER, Bark Patch. **Aleurodiscus** spp., eastern states; **Felt Fungus, Septobasidium** spp., NC to FL.

CANKER, Branch, Trunk. **Dothiorella fraxinicola**, IA, KS, NE; **Sphaeropsis** sp., widespread; **Nectria cinnabarina**; **N. coccinea**, Northeast.

CANKER, Coin. **Phyctema vagabunda**, MT, Ontario, Canada.

CANKER; DIEBACK. **Cytospora annularis**, north central states; **Diplodia infuscans**, Northeast; **Cytophoma pruinosa** and **Fusicoccum** sp., may be secondary, Northeast.

DECLINE. **Togninia fraxinopennsylvanica**, Anamorph, **Phaeoacremonium mortoniae**, CA.

DODDER. **Cuscuta** sp.

LEAF SPOT. **Cercospora fraxinites**, AL, FL, LA, TX; **Cylindrosporium fraxini** (including reports of *Marssonina fraxini*, *Piggotia fraxini*), widespread;

Mycosphaerella effigurata, widespread; **M. fraxinicola**; **Actinopelte dryina**; **Cytospora** sp., VA.

MEASLES. **Togninia fraxinopennsylvanica** Anamorph, **Phaeoacremonium mortoniae**, CA.

MISTLETOE. **Phoradendron serotinum (flavescens)**, south central to Pacific states; **Viscum album**, CA, on *Fraxinus velutina* (Arizona Ash).

MISTLETOE, European. **Viscum album** on Arizona ash, CA.

NEMATODE. **Meloidogyne** spp., AZ, MD, OK.

NEMATODE, Dagger. **Xiphinema americanum**, SD on Green Ash.

POWDERY MILDEW. **Phyllactinia corylea**, northeastern, central, Pacific Coast states. **Uncinula circinata**, IA; **Microsphaera alni**, IL.

ROT, Collar. **Helicobasidium purpureum**, TX.

ROT, Root. **Phymatotrichum omnivorum**, AZ, TX.

ROT, Sapwood. **Lentinus tigrinus**, MS.

ROT, White Mottled Heart, Stem. **Fomes fraxinophilus**, eastern and central states to Great Plains; many other species of *Fomes*.

ROT, Wood. **Daldinia concentrica**, cosmopolitan; **Daedalea** spp.; **Ganoderma lucidum**, LA, MS; **Polyporus** spp.; **Poria** spp.; **Schizophyllum commune**, cosmopolitan; **Trametes** spp.

RUST. **Puccinia sparganioides** (0, I), general east of Great Plains; II, III on marsh grass (*Spartina* spp.).

VIRUS. **Ash Ring Spot**, NY; **Tobacco Mosaic Virus**, MA, NY; **Tobacco Ring Spot Virus**, NY.

WILT. **Verticillium albo-atrum**, CO; **V. dahliae**, OR.

Ash rust is epidemic in New England in many seasons, causing defoliation and sometimes death of trees. Anthracnose and leaf spots may be important in a wet season.

ASH, MORAINE (*Fraxinus holotriocha*)

VIRUS. **Tobacco Mosaic**, NY; **Tobacco Ringspot**, NY.

AUCUBA (Gold-Dust Tree)

ANTHRACNOSE; Leaf Spot. **Gloeosporium** sp. (*Glomerella cingulata*), NJ, PA, SC, WA. Probably includes reports of *Colletotrichum pollaci*, MS, NJ.

LEAF SPOT. **Pestalotia aucubae**, secondary; **Phyllosticta aucubae**, CA, MS, SC.

ROT, Root. **Phytophthora cinnamomi**, **P. citricola**, NC.

VIRUS. **Tobacco Mosaic**, Aucuba strain.

WILT. **Verticillium albo-atrum**, NJ.

AVOCADO (*Persea americana*)

ANTHRACNOSE; Leaf and Fruit Spot; Black Rot. **Glomerella cingulata**, Mexico, general; **Colletotrichum gloeosporioides**, Mexico.

BACTERIAL CANKER. **Xanthomonas campestris**, CA.

BACTERIAL Crown Gall. **Agrobacterium tumefaciens**, CA.

BACTERIAL Fruit Blast. **Pseudomonas syringae**, CA.

BLACK MILDEW. **Irene perseae**, FL.

BLACK STREAK. No causal agent has been identified, FL.

BLIGHT, Seedling. **Sclerotium rolfsii**, FL; **Phytophthora palmivora**, FL.
BLOTCH, Fruit Spot. **Cercospora perseae**, FL; **C. purpurea**.
CANKER, Branch; Dothiorella Fruit Rot. **Botryosphaeria ribis** var. **chromogena**, CA; **Neofusicoccum austral**, **N. luteum**, **and N. parvum**, CA.
LEAF SPOT. **Phyllosticta micropuncta**, AL, FL.
LEAF SPOT; Algal; Green Scurf. **Cephaleuros virescens**, FL.
LEAF SPOT; Fruit Spot; Seedling Blight. **Pestalotia** spp., general.
LEAF SPOT; Smudgy Spot on Twigs. **Helminthosporium** sp., CA.
NEMATODE, Burrowing. **Radopholus similis**, FL.
NEMATODE, Dagger. **Xiphinema americanum**.
NEMATODE, Meadow. **Pratylenchus brachyurus**; **P. vulnus**.
NEMATODE, Root Knot. **Meloidogyne** sp., resistant to.
NEMATODE, Stubby Root. **Trichodorus christiei**.
NONPARASITIC. **Carapace Spot**. Abrasion of young fruits.
Dieback. Copper deficiency.
End Spot. Dessication of young fruits; overmaturity.
Little Leaf; Rosette. Zinc deficiency.
Melanorhiza. Defective drainage and aeration.
Mottle Leaf. Nutritional deficiency.
Tipburn. Sometimes salt accumulation in poorly drained soil.
ROT, Blue Mold. **Penicillium expansum**, CA.
ROT, Collar; Trunk Canker. **Phytophthora cactorum** CA; **P. parasitica** FL; **Sclerotinia sclerotiorum**, CA.
ROT, Fruit. **Diplodia theobromae**, FL; **Phomopsis** sp., FL, TX; **Phytophthora citrophthora**, CA; **P. citricola**, CA; **Rhizopus nigricans**; **Alternaria** sp.; **Fusarium** sp.
ROT, Postharvest Stem. **Pestalotiopsis clavispora**, Chile.
ROT, Root. **Armillaria mellea**, CA; **Clitocybe tabescens**, FL; **Phymatotrichum omnivorum**, TX; Dematophora Rot, **Rosellinia necatrix**.
ROT, Root; Decline. **Phytophthora cinnamomi**, CA; **Pythium** spp., CA.
ROT, Seed and Root. **Rhizoctonia solani**, CA.
SPOT ANTHRACNOSE; Scab. **Sphaceloma perseae**, FL, TX.
VIROID. **Avocado Sunblotch**, Michoacan, Mexico.
VIRUS. **Avocado Sun Blotch**, CA; **Avocado 3 Alphacryptovirus**., CA, black streak.
WILT. **Verticillium albo-atrum**, CA, FL; **Raffaelea lauricola**, FL.
WILT, Laurel. **Raffaelea lauricola**; **R. canadensis**, CA, FL.
WILT, Root. **Neonectria radicicola**, CA.

Avocado scab may cause heavy damage in susceptible varieties of West Indian stock. Cercospora blotch attacks both leaves and fruit. Decline, root rot, is most serious in wet soils, killing trees if they are waterlogged 6–8 days.

AZALEA (*Rhododendron*)
ANTHRACNOSE. **Colletotrichum** sp., LA.
BLIGHT, Bud and Twig. **Briosia azaleae** (*Pycnostysanus*), MA, NH, NJ, NC.
BLIGHT, Cutting. **Cylindrocladium scoparium**, AL, FL, NY, OH; **C. floridanum**.

BLIGHT, Flower; Seedling. **Botrytis cinerea**, often after frost, cosmopolitan; **Alternaria tenuis**; **Cladosporium herbarum**; **Epicoccum purpurascens**.

BLIGHT, Foliar. **Phytophthora foliarum**, TN.

BLIGHT, Petal; Flower Spot. **Ovulinia azaleae**, AL, CA, FL, GA, LA, MD, MS, NC, PA, SC, TX, VA; also NY, NJ in greenhouses.

BLIGHT, Shoot. **Monilinia azaleae**, GA, MA, NY.

BLIGHT, Shoot and Stem. Phytophthora citrophthora, P. citricola, P. nicotianae var. parasitica, FL, OH.

BLIGHT, Thread. **Pellicularia koleroga**, LA.

CANKER, Stem. **Gloeosporium** sp., OR. **Phomopsis** sp., SC.

DAMPING-OFF, Leaf Blight. **Rhizoctonia solani**, cosmopolitan.

DODDER. **Cuscuta** sp., FL, SC.

GALL, Leaf; Shoot Hypertrophy. **Exobasidium vaccinii**, general; **E. burtii**, **Yellow Leaf Spot**, ID, NJ, OR, WA; **Synchytrium vaccinii**, NJ.

LEAF SCORCH; Angular Leaf Spot. **Septoria azaleae**, widespread; severe in CA.

LEAF SPOT. **Cercospora rhododendri**, MD; **Pestalotia** spp., general but secondary; **Phyllosticta** sp.; **Ramularia angustata**, MS, NY; **Colletotrichum azaleae**, FL; **Septoria solitaria**, CA, OR, TX; Tar Spot, **Melasmia menziesii**, WA; **Corynespora cassiicola**, GA, FL.

NONPARASITIC. **Chlorosis**. Usually iron deficiency, general in alkaline soils, sometimes defective drainage.

NEMATODE, Leaf. **Aphelenchoides fragariae**, FL.

NEMATODE, Root Knot. **Meloidogyne incognita**.

NEMATODE, Spiral. **Helicotylenchus nannus**; **Rotylenchus robustus**.

NEMATODE, Stubby Root. **Paratrichodorus christiei**; **Trichodorus primitivus**.

NEMATODE, Stunt. **Tylenchorhynchus claytoni**.

POWDERY MILDEW. **Erysiphe azaleae**, LA; **Erysiphe polygoni**, CA, NJ; **Microsphaera alni**, GA, NJ, NY, PA, RI, VA.

ROT, Root. **Armillaria mellea**, CA, NJ, WA; **Clitocybe tabescens**, FL; **Phymatotrichum omnivorum**, TX; **Pythium** spp.; **Phytophthora lateralis**, NC.

ROT, Root and Stem; Wilt. **Phytophthora cinnamomi**, AL, MD, MO.

RUST. **Pucciniastrum myrtilli** (II, III), ME to FL, TX; 0, I on hemlock.

VIRUS. **Tomato Spotted Wilt**, PA.

WILT. **Cylindrocarpon radicicola**, MA.

Azalea flower spot or petal blight devastates azalea blooms in the South in rainy or foggy weather starting when buds show color. Bud and twig blight has killed some shrubs in Massachusetts. Leaf galls are unsightly but not too serious. Leaf scorch may be prevalent in a rainy season. Powdery mildew is common on deciduous azaleas in late summer.

AZARA

ROT, Stem. **Sclerotium rolfsii**, CA.

B

BALD CYPRESS (*Taxodium*)

BLIGHT, Twig. **Pestalotia funerea**, TX.

CANKER, Felt Fungus. **Septobasidium** spp., LA.

ROT, Brown Pocket Heart. **Fomes geotropus**, cause of "pecky cypress," FL to LA; **F. extensus**; **Ganoderma applanatum**.

ROT, Root. **Phymatotrichum omnivorum**, TX.

ROT, Wood. **Lenzites** spp., **Polyporus** spp., **Poria** spp.

BAMBOO (*Bambusa*, *Phyllostachys*)

BLACK MILDEW. **Meliola tenuis**.

BLIGHT, Tip. **Diplodia bambusae**, LA, TX.

LEAF MOLD. **Cladosporium gramineum**, OR, SC.

LEAF SPOT. **Helminthosporium** sp., FL; **Mycosphaerella** sp., CA; Culm Spot, **Selenophoma donacis**, CA.

NEMATODE, Burrowing. **Radopholus similis**, FL.

RUST. **Kweilingia divina**, CA, HI, Brazil, Colombia; **Puccinia melanocephala** (II, III), FL, GA, MS, TX; **Uredo ignava** (II), FL, GA; **Uromyces costaricensis**, FL on Wild Bamboo (*Lasiacis divaricata*).

SMUT. **Ustilago shiraiana**, CA, FL, LA, TX.

VIRUS. **Bamboo Mosaic**, CA; **Sugarcane Mosaic Virus**, LA, on *Arundinaria gigantea*.

BANANA (*Musa*)

ANTHRACNOSE; Fruit Rot. **Gloeosporium musarum**, FL, TX.

BACTERIAL Leaf Blight. **Pseudomonas solanacearum**, FL.

BLACK SIGATOKA. **Mycosphaerella fijiensis**, FL; Grand Bahama Island, Puerto Rico.

FUZZY PEDICEL. **Sporothrix** spp., **Fusarium pseudocircinatum**, FL.

LEAF SPOT, Black. **Deightoniella**, GA.

NEMATODE, Burrowing. **Radopholus similis**, FL, LA.

NEMATODE, Lesion. **Pratylenchus musicola**, FL.

NEMATODE, Root Knot. **Meloidogyne** spp., FL, TX.

R.K. Horst, *Field Manual of Diseases on Trees and Shrubs*,
DOI 10.1007/978-94-007-5980-0_5, © Springer Science+Business Media Dordrecht 2013

Rot, Root. **Clitocybe tabescens**, FL; **Ceratobasidium**, GA.
Virus. **Cucumber Mosaic**, FL.
Wilt. **Fusarium oxysporum** f. sp. **cubense**, FL.

BANANA SHRUB (*Michelia figo*)
Powdery Mildew. **Oidium** sp., LA.

BARBERRY (*Berberis*)
Bacterial Leaf Spot. **Pseudomonas berberidis**, general.
Blight, Gray Mold. **Botrytis cinerea**, CA, MO.
Canker; Dieback. **Botryosphaeria ribis**, FL.
Damping-Off. **Pythium debaryanum**, CA.
Leaf Spot. **Gloeosporium berberidis**, CT, MA, MN, OH, WI; **Phyllosticta berberidis**, KY.
Nematode, Root Knot. **Meloidogyne hapla**.
Nematode, Spiral. **Rotylenchus buxophilus**.
Powdery Mildew. **Erysiphe polygoni**; **Phyllactinia corylea**, MA, VT.
Rot, Heart. **Poria punctata**, MD.
Rot, Root. **Phymatotrichum omnivorum**, TX.
Rust. **Cumminsiella sanguinea** (0, I, II, III), OR; **Puccinia koeleriae** (0, I); II, III on Koehleria; **P. montanensis** (0, I); II, III on grasses.
Rust, Stripe. **Puccinia striiformis**, MN.
Rust, Wheat, **Puccinia graminis** (0, I), general; II, III on cereals and grasses.
Virus. **Tomato Spotted Wilt**, PA.
Wilt. **Verticillium albo-atrum**, CT to VA, IL, MI.

All interstate movement of barberry is under quarantine because of the wheat rust, but resistant cultivars may be shipped under permit. Common barberry is eradicated near wheat fields; Japanese barberry is resistant to rust.

BAUHINIA (Orchid-Tree, Mountain Ebony)
Leaf Spot. **Colletotrichum** sp., TX; **Phyllosticta** sp., FL.
Powdery Mildew. **Microsphaera diffusa**, MD.

BAYBERRY (*Myrica carolinensis*)
Leaf Spot. **Mycosphaerella myricae**, GA, MS; **Phyllosticta myricae**, NY.
Rust. **Gymnosporangium ellisii** (0, I), MA to NY, VA; III on *Chamaecyparis*.
Virus. **Bayberry Yellows**, NJ.

BEARBERRY (*Arctostaphylos uva-ursi*)
Black Mildew. **Asterina gaultheriae**, WI; **A. conglobata**, ME.
Gall, Leaf; Red Leaf Spot. **Exobasidium vaccinii**, CO, ID, MA, MT, NJ, NY, VT, WI.
Leaf Spot. **Cercospora gaultheriae**, WI.
Rust. **Chrysomyxa arctostaphyli** (III), AK, CO, MT, UT, WI; **Puccinia sparsum** (II, III), WI.

BEAUTY-BUSH (*Kolkwitzia*)
Leaf Spot. **Cercospora kolkwitziae**, AL, OK.

BEECH (*Fagus*)

BROOMRAPE. **Conophilis americana**; **Epifagus virginiana**, beechdrops.

CANKER, Beech Bark Disease. **Nectria coccinea** var. **faginata**, associated with woolly beech scale, destructive northern New England and NY, OH; **N. galligena**, New England, MI, NY, NC, PA, WV; **Cryptococcus fagesuiga**, OH, VA.

CANKER, Bleeding. **Phytophthora cactorum**. MA, NY, RI.

CANKER; DIEBACK. **Cytospora** spp.; **Strumella coryneoides**, New England; **Phomopsis** sp., MI, NY, PA.

CANKER, Felt Fungus. **Septobasidium** spp.; **S. cokeri**, **S. curtisii**, FL, NC.

CANKER, Perennial. **Endothia gyrosa**, NY.

LEAF SPOT. **Cercospora** sp., NY; **Gloeosporium fagi**, CT to WI; **Microstroma** sp., IL, NJ; **Phyllosticta faginea**, MA to WV.

MISTLETOE. **Phoradendron serotinum** (**flavescens**), occasional VT to IN and southward.

NONPARASITIC. **Leaf Scorch**, common in Northeast but cause unknown, associated with high temperature and water deficiency.

Leaf Mottle, cause unknown, virus suspected, CT, NJ, NY, PA.

POWDERY MILDEW. **Microsphaera alni**, MA to AL, WI; **Phyllactinia corylea**, New England to AL, IL, WI.

ROT, Heart, Butt, Wound. **Fomes** spp.; **Ganoderma applanatum**; **Polyporus** spp.; **Schizophyllum commune**; **Stereum** spp.

ROT, Mushroom Root. **Armillaria mellea**, CT, NJ, NY, OH.

ROT, Sapwood. **Hericium** spp.; **Ustulina deusta**.

ROT, Wood. **Daedalea** spp.; **Daldinia** spp.; **Lenzites betulina**, widespread.

SOOTY MOLD. **Scorias spongospora**, MA to AL, MO.

BIGNONIA (*Crossvine*)

BLACK MILDEW. **Asterina bignoniae**, LA; **Meliola bidentata**, Gulf states; **M. furcata**, FL.

LEAF SPOT. **Cercospora capreolata**, AL, MS; **Leptostromella bignoniae**, TN.

NEMATODE, Root Knot. **Meloidogyne** sp.

SOOTY MOLD. **Capnodium elongatum**, AL, MS.

SPOT ANTHRACNOSE. **Sphaceloma** sp., LA.

BIRCH (*Betula*)

ANTHRACNOSE. **Gloeosporium** sp.; **Glomerella cingulata**, VA, on Eur. White Birch.

CANKER, Bark Patch. **Aleurodiscus oakesii**, VT; **Solenia ochracea**, cosmopolitan.

CANKER, Bleeding, **Phytophthora cactorum**, NJ.

CANKER, Trunk. **Hymenochaete agglutinans**, MI, PA; **Nectria coccinea**; **N. galligena**; Twig Canker, **N. cinnabarina**.

DIEBACK. **Melanconis stilbostoma**, MA to IN on white birch; other species; **Cylindrosporium orthosporum**.

LEAF BLISTER. **Taphrina americana**, Witches' Broom, NH, VT, WI; **T. carnea**, ME, NH; **T. flava**, ME, MA, NH, WI.

LEAF SPOT. **Gloesporium betularum**; **G. betulaeluteae**, NY, PA; **Cylindrosporium betulae** DE, NY, WI; **Phyllosticta betulinum**; **Septoria betulicola**; **S. betulae**.

MISTLETOE. **Phoradendron serotinum** (**flavescens**), FL, IN, TX; **Viscum album**, CA, on *Betula verrucosa* (Weeping White Birch).

POWDERY MILDEW. **Microsphaera alni**, MA, NH, PA; **Phyllactinia corylea**, widespread.

ROT, Heart. **Fomes** spp.; **Stereum purpureum**; **Polyporus hispidus**.

ROT, Root. **Armillaria mellea**; **Phytophthora cinnamomi**, of seedlings; **Phymatotrichum omnivorum**, **TX**.

ROT, Sapwood. **Polyporus betulinus**, general on gray and paper birches; **P. gilvus**; **Steccherinum** spp.

ROT, Wood. **Polyporus** spp., **Stereum** spp.; **Daedalea unicolor** and other spp.

ROT, Wound. **Schizophyllum commune**; **Pleurotus serotinus**.

RUST. **Melampsoridium betulinum** (II, III); 0, I on larch.

VIRUS. Dieback, vein clearing, New England.

BISCHOFIA

LEAF SPOT, Algal. **Cephaleuros virescens**, FL.

BITTERSWEET (Celastrus)

BACTERIAL Crown Gall. **Agrobacterium tumefaciens**, CT.

CANKER; DIEBACK. **Glomerella cingulata**, NH, NC.

LEAF SPOT. **Ascochyta** sp., CT; **Asteria celastri**, **KS**, **ME**, **MI**; **Cercospora melanochaeta**, IA, KS, NE, ND; **Marssonina** sp., NY; **Phyllosticta celastri**, IL, KS, MA, NY, WV; **Ramularia celastri**, VT to MS, ND, TX.

POWDERY MILDEW. **Microsphaera alni**, WI; **Phyllactinia corylea**, ME to VA, SD, TX.

ROT, Root. **Phymatotrichum omnivorum**, TX.

BIXA (*Annato-Tree*)

LEAF SPOT. **Cercospora bixae**; **Phyllosticta bixae**.

LEAF SPOT, Algal. **Cephaleuros virescens**, FL.

BLACKBERRY (*Rubus*)

ANTHRACNOSE. **Glomerella cingulata**, Mexico.

BACTERIAL Crown Gall. **Agrobacterium tumefaciens**, general; Cane Gall, **A. rubi**, NY, OR, PA, WA, WI; Hairy Root, **A. rhizogenes**, OR.

BLIGHT, Cane. **Gnomonia rubi**, ME, MD, NY, PA, VT; **Leptosphaeria coniothyrium**, NY to NC, TX, WI, Pacific Northwest.

BLIGHT, Leaf. **Colletotrichum acutatum**, Brazil.

BLIGHT, Spur. **Didymella applanata**, VA, WI.

BLIGHT, Stamen; Dry Berry. **Hapalosphaeria deformans**, OR, WA.

BLIGHT, Thread. **Pellicularia koleroga**, LA.

BLOTCH, Sooty. **Gloeodes pomigena**, MD to NC, IN, TX; **Geastrumia polystigmatis**, NC.

CANKER. **Glomerella cingulata**, MD, VA; **Phomopsis** sp., WA; **Botryosphaeria obtusa**, OH; **B. dothidea**, MD; **Leptosphaeria coniothyrium**, OH; **Gnomonia rubi**, OH.

CANKER, Cane Spot. **Ascospora ruborum**, WA.

Downy Mildew. **Peronospora rubi**, MD, WI; **P. sparsa**, Mexico.
Fruit Spot; Flyspeck. **Leptothyrium pomi**, PA to NC, IL.
Leaf Spot. **Mycosphaerella confusa**, VA to FL, IL, IN, TX; **M. rubi**, NC, WI; **Pezizella oenotherae**, also fruit rot, MD, OH, VA; **Cylindrosporium rubi**, NC, TX; **Phyllosticta** spp., FL, IL, NH; **Septoria darrowi**; **S. rubi**, general.
Leaf Spot, Algal. **Cephaleuros virescens**, AR, FL, LA.
Powdery Mildew. **Sphaerotheca macularis**, CT to MD, IL, MN, Pacific Northwest; **Phyllactinia corylea**, MI.
Rosette, Double Blossom. **Cercosporella rubi**, NY to FL, CA, TX.
Rot. **Botrytis** spp., SC.
Rot, Gray Mold. **Botrytis cinerea**, general on fruit, bud, shoot.
Rot, Root. **Armillaria mellea**, TX, WA; **Phymatotrichum omnivorum**, TX; **Phytophthora dreschleri**, NC; **Rhizoctonia solani**.
Rot, White Root. **Corticium galactinum**, near apple trees, AR, MD, VA.
Rust. **Phragmidium violaceum**, CA.
Rust, Orange. **Gymnoconia peckiana** (0, I, III), ME to GA and west to Pacific; **Kunkelia nitens** (I), general, more common in South; **Phragmidium violaceum**.
Rust, Yellow Cane. **Kuehneola uredinis** (0, I, II, III), ME to FL, LA, WI.
Spot Anthracnose. **Elsinoë veneta**, general.
Virus. **Blackberry Dwarfing**; **Blackberry Mosaic**; **Blackberry Variegation**; **Loganberry Dwarf**, OR; **Raspberry Beta-Leaf Curl**; **Red Raspberry Mosaic**, MA to VA, IA, WI, Pacific Northwest; **Raspberry Ringspot**; **Raspberry Streak**, OH, PA, WA; **Tobacco Streak**, OR, WA; **Wineberry Latent**; **Beet Pseudo-Yellows**, AR, NC, SC; **Marafivirus**, MS; **Impatiens Necrotic Spot**, AL, Southern U.S; **Tobacco Ringspot**, **Tomato Ringspot**, **Raspberry Bushy Dwarf**, AL; **Black Raspberry Necrosis**, TN; **Blackberry Chlorotic Ringspot**, OR, WA.
Wilt. **Verticillium albo-atrum**, CA, MN, NY, WA.

Sanitation is the best approach to home garden blackberry diseases. Plants with crown gall, orange rust, or virus diseases should be removed and burned, replanting with clean stock.

BLADDER-SENNA (*Colutea*)
Blight, Twig. **Diplodia coluteae**, PA.
Powdery Mildew. **Erysiphe polygoni**.
Rot, Root. **Ganoderma** sp., OK; **Phytophthora cactorum**, MO.
Rust. **Uromyces colutea** (II, III), KS.

BLEEDING-HEART VINE (*Clerodendrin*)
Virus. **Zonate Ring Spot**, FL.

BLUEBERRY (*Vaccinium*)
Anthracnose. **Gloeosporium** sp., FL, NJ.
Bacterial Canker. **Pseudomonas** sp., OR; **Xanthomonas campestris** pv. **manihotis** (also leaf spot), FL; **X. campestris** pv. **fici** (also leaf spot), FL.

BACTERIAL Crown Gall. **Agrobacterium tumefaciens**, MA, MI, WA.

BACTERIAL Leaf Scorch, **Xylella fastidiosa**, FL.

BLACK MILDEW, **Meliola nidulans**, AL, GA.

BLIGHT, Blossom, Fruit, Twig. **Botrytis cinerea**, CA, ME, MA, NJ, OR, WA; **Diaporthe vaccinii**, ME, MA, NJ.

BLIGHT, Stem. **Botryosphaeria dothidea** (*B. ribis*), NC; **Phomopsis vaccinii**, NC; **Lasiodiplodia theobromae**, FL; **Fusarium** sp., Argentina; **Calonectria colhounii**, MI.

BLIGHT, Twig. **Nigrospora sphaerica**, Argentina.

CANKER, Blight of Stem, MI. **Glomerella cingulata**, **Alternaria** sp., **Sordaria** sp., **Epicoccum** sp., **Tympanis** sp., **Papulospora** sp., **Nectria cinnabarina**, **Coniothyrium** sp., **Pestalotia** sp., **Verticillium** sp., **Godronia cassandrae** f. sp. **vaccinii**, **Phoma** sp., **Dendrophoma** sp., **Diaporthe** (**Phomopsis**) **vaccinii**, **Melanospora** sp., **Pullularia** spp., **Coryneum microstictum**, **Sphaeronema** sp., **Fusarium** spp., **Bispora** sp., **Botrytis cinerea**, **Cephalosporium** sp., **Cylindrocarpon** sp., **Pyrenochaeta** sp.

CANKER, Cane. **Physalospora corticis**, AL, FL, MS, NC; **Godronia cassandrae**, ME, MI; **Botryosphaeria corticis**, AL, FL, GA, MS, NC; **Coryneum microstictum**, MA, MI.

CANKER, Twig. **Dothidella vacciniicola**, NC; **Fusicoccum putrefaciens**, ME, MA, MI, **Phomopsis vaccinii**, MI; **Sporocadus lichenicola**, OR.

DODDER. **Cuscuta** sp., FL.

GALL, Crown and Root. **Agrobacterium rubi** and **A. rhizogenes**, Argentina.

GALL, Red Leaf. **Exobasidium vaccinii**, also fruit green spot, general; **Synchytrium vaccinii**, ME, MS.

LEAF RUST. **Thekopsora minima**, MI, Mexico; **Pucciniastrum corymbosum**, HI.

LEAF SPOT. **Gloeocereocospora inconspicua**, GA, MD, NC; **Phyllosticta** sp., AL, GA; **Phyllostictina vaccinii**, GA, MD, MI, NC; **Piggotia vaccinii**, WI; **Discohainesia oenotherae**, NJ, NC; **Ramularia vaccinii**, FL, MD, MI, NJ, NY; **Septoria albopunctata**; **Alternaria tenuissima**, NC; **Cristulariella pyramudates**, MD; **Nigrospora sphaerica**, Argentina.

LEAF SPOT, Double Spot. **Dothichiza caroliniana**, GA, NC.

LEAF SPOT, Tar. **Thytisma vaccinii**, widespread.

NEMATODE. **Helicotylenchus** sp.; **Hemicycliophora similis**; **Merlinius joctus**; **Xiphinema Americanum**; **Paratrichodorus renifer**, WA, British Columbia, Canada.

POWDERY MILDEW. **Microsphaera alni** var. **vaccinii**, widespread.

ROT, Berry; Dieback. **Alternaria** sp., MA, NJ, NC; **Glomerella cingulata**, NJ, NY; **Phomopsis vaccinii**, NC.

ROT, Brown; Mummy Berry. **Monilinia vaccinii-corymbosi**, IN, ME, MA, MI, NJ, NY, NC, PA; **Sclerotinia sclerotiorum**, Argentina.

ROT, Root. **Phytophthora cinnamomi**, AR, NJ, VA; **P. nicotainae**, VA; **P. palmivora**, VA; **Fusarium proliferatum**, Argentina.

Rust, Witches' Broom. **Pucciniastrum goeppertianum** (III) ME to MN; 0, I on fir; Leaf, **P. myrtilli** (II, III) ME to PA, WI; 0, I, spruce.

Virus. **Blueberry Necrotic Ring Spot**; **Blueberry Red Ring Spot**; **Blueberry Stunt**, MA, MI, NJ, NY, NC; **Blueberry Shoestring**, MI, NJ, WA; **Blueberry Leaf Mottle**, MI; **Blueberry Necrotic Shock**, OR, WA; **Blueberry Scorch**, CT, MA, OR, WA, British Columbia, Canada; **Tobacco Ring Spot**, IL; **Peach Rosette Mosaic**, MI.

BOEHMERIA

Nematode. **Paratylenchus elachistus**, FL.

BOTTLE-BRUSH (*Callistemon*)

Leaf Spot. **Cylindrocladium colhounii**, FL, HI, LA, NC, OR, SC; **C. clavatum**, FL.

Rot, Root. **Armillaria mellea**, CA.

BOUGAINVILLEA (*Buginvillaea*)

Bacterial Leaf Spot. **Pseudomonas andropogonis**, FL; **Burkholderia andropogonis**, Canada.

Blight, Foliage. **Phytophthora parasitica**, FL.

Leaf Spot. **Cladosporium arthrinioides**, TX; **Cercospora bougainvilleae**, FL; **Passalora bougainvilleae**, Mexico.

Virus. **Mosaic**. Undetermined, FL.

BOUVARDIA

Nematode, Leaf. **Aphelenchoides fragariae**.

Nematode, Root Knot. **Meloidogyne** sp., NY.

Rust. **Puccinia bouvardiae** (0, I, III), AZ.

BOYSENBERRY

Subject to most blackberry diseases: anthracnose, crown gall, cane gall, cane canker, dieback, leaf spots, mosaic.

BOXELDER (*Acer negundo*)

Anthracnose. **Gloeosporium apocryptum**, widespread.

Bacterial Leaf Spot. **Pseudomonas aceris**, CA.

Blight Leaf. **Coryneum negundinis**, MO; on twigs, ME.

Blight, Twig. **Coniothyrium negundinis**, IL, OK; **Coryneum negundinis**, ME. **Nectria cinnabarina.**

Canker, Felt Fungus. **Septobasidium** spp., NC.

Leaf Spot. **Alternaria** sp.; **Ascochyta negundinis**, IL, NC: **Cercosporanegundinis**, KS, NE, WI; **Piggotia negundinis**, WI; **Phyllosticta minima**, general; **P. negundinis**, ME to AL, TX, WI; **Septoria aceris**, general; **S. negundinis**; **Cylindrosporium negundinis**; **Cristulariella pyramidalis**.

Leaf Spot, Tar. **Rhytisma punctatum**, CA, NY; **R. acerinum**, OR, WA.

Powdery Mildew. **Phyllactinia corylea**, SD; **Microsphaera alni**, **Uncinula circinata**; **Sawadaea bicornis**, WI.

ROT, Root. **Phymatotrichum omnivorum**, CA, TX; **Helicobasidium purpureum**, TX.

WILT. **Verticillium albo-atrum**, occasional.

BOXWOOD (*Buxus*)

BLIGHT, Box. **Cylindrocladium pseudonaviculatum**, NC.

BLIGHT, Leaf Cast. **Hyponectria buxi**, general; **Verticillium buxi**, cosmopolitan on dead leaves, often associated with *Hyponectria*.

BLIGHT, Leaf Tip. **Phoma conidiogena**, MD, NJ, NY, OK.

CANKER, "Nectria"; Leaf Blight. **Volutella buxi**, general; considered imperfect stage of *Pseudonectria rouselliana* but unconfirmed connection.

CANKER; DIEBACK. **Fusarium buxicola** (*Nectria desmazierii*), AL, MD, PA; **F. lateritium** (see *Gibberella baccata*), secondary, MD, SC, VA.

LEAF SPOT. **Phyllosticta auerswaldii**, MD, MA, NJ, NY, VA, WA; **Macrophoma candollei**, prominent, general on dead leaves following winter injury or disease.

NEMATODE, Dagger. **Xiphinema americanum**.

NEMATODE, Lesion. **Pratylenchus pratensis**; **P. vulnus**, NC.

NEMATODE, Root Knot. **Meloidogyne hapla**; **M. incognita**; **M. incognita acrita**; **M. javanica**; **M. arenaria**, NC.

NEMATODE, Spiral. **Helicotylenchus nannus**; **Rotylenchus buxophilus**.

NEMATODE, Stubby Root. **Trichodorus** sp.

NONPARASITIC. **Freezing**. Ice standing on stems causes bark to slough off and branches die back for months thereafter.

Sunscald. Injury in late winter or early spring when covering is removed in bright sun or high wind.

ROT, Root. **Armillaria mellea**, NJ; **Fusarium oxysporum** and **F. solani**, MD, perhaps secondary; **Phymatotrichum omnivorum**, TX; **Phytophthora parasitica**, MD, NC; **P. cactorum**, NC; **P. citrophthora**, NC; **Pythium** sp., MA; **Rhizoctonia solani**, cosmopolitan after nematode injury.

ROT, Heart; Trunk. **Ganoderma lucidum**, VA; **Fomes igniarius**, VA; **Poria punctata**, VA.

ROT, Root; Decline. **Paecilomyces buxi** and some **Fusaria**, VA.

VIRUS. **Variegation**, cause unknown, MD, NY, VA.

Salmon-pink pustules appearing on backs of leaves, along twigs, and on main stems are indications of *Volutella buxi*, controlled by thorough cleaning, then spraying. Much of the dieback, bronzing, general unhealth of boxwood is due to nematodes, especially spiral and root lesion or meadow.

BRICKELLIA (Brickle-Bush)

DODDER, **Cuscuta exaltata**, TX.

LEAF SPOT. **Cercospora coleosanthi**, CA, CO.

ROT, Root. **Phymatotrichum omnivorum**, TX.

RUST. **Aecidium arcularium** (0, I), AZ, CO, NM; **Coleosporium aridum** (II), CA; **Puccinia kuhniae**, AZ, FL; **P. subdecora** (0, I, II, III), AZ, CO, NM, UT; **Uredo arida** (II), CA.

BROOM (*Cytisus*)

CANKER. **Diaporthe** spp., NJ, NY; **Dothidea tetraspora**, CA; **Gloeosporium** sp., NJ; **Nectria coccinea**, OR; **Pestalotia polychaeta**, CA; **Phomopsis** sp., MA, NJ; **Physalospora obtusa**, AL. Many are secondary.

NEMATODE, Lesion. **Pratylenchus pratensis**.

ROT, Root. **Thielaviopsis basicola**, WI.

BROOM, SPANISH (*Spartium*)

CANKER. **Diplodia sarothamni**, CA; **Pestalotia polychaeta**, CA; **Phomopsis sarothamni**, CA. All possibly secondary infections.

BROUSSONETIA (Paper-Mulberry)

CANKER; DIEBACK. **Nectria cinnabarina**, AL, NY; **Fusarium solani**, OH.

MISTLETOE. **Phoradendron serotinum (flavescens)**, TX.

NEMATODE, Root Knot. **Meloidogyne** sp., MD.

ROT, Root. **Phymatotrichum omnivorum**, TX; **Armillaria mellea**, CA.

BRUNFELSIA

BACTERIAL Canker; Vascular. **Clavibacter michiganense**, WY.

BUCKEYE

See Horse-Chestnut.

BUCKLEYA

RUST. **Cronartium comandrae** (II, III), TN; 0, I on pine.

BUCKTHORN (Rhamnus)

LEAF, Black. **Phytophthora ramorum**, CA.

LEAF SPOT. **Cercospora aeruginosa**, MO, NE, SC; **C. Bacilligera**; **C. rhamni**, LA, NE, NJ, TX, WI; **Cylindrosporium rhamni**, ID; **Marssonina rhamni**, WA; **Phaeosphaerella rhamni**, CA; **Phyllosticta rhamnigena**; **Septoria blasdalei**, CA, ID, OR, TX.

POWDERY MILDEW. **Microsphaera alni**, WI.

ROT, Heart. **Daedalea unicolor**, WA; **Fomes igniarius**, ID.

ROT, Root. **Phymatotrichum omnivorum**, TX.

RUST. **Puccinia coronata** (0, I), widespread; II, III on grasses, cereals, widespread; **P. mesneriana** (III), CA.

SOOTY MOLD. **Capnodium** sp., CA.

BUCKWHEAT-TREE (*Cliftonia*)

BLACK MILDEW. **Morenoella cliftoniae**, MS.

LEAF SPOT. **Coccomyces** sp.; **Pestalotia cliftoniae**, MS.

BUDDLEIA (Butterfly-Bush)

CANKER, Stem. **Phomopsis buddleiae**, AZ.

NEMATODE, Root Knot. **Meloidogyne** spp., AL, MS, TX.

ROT, Root. **Phymatotrichum omnivorum**, TX.

SCAB. **Cladosporium heugelinianum**, DE, VA.

VIRUS, Line Spot. **Tomato Ringspot**, SC.

BUFFALOBERRY (Shepherdia)

Damping-Off. **Pythium ultimum**, NE; **Rhizoctonia solani**, NE.

Leaf Spot. **Septoria shepherdiae**, AK, ID, MT, UT, WI; **Cylindrosporium** sp., Great Plains.

Powdery Mildew. **Phyllactinia corylea**, UT; **Sphaerotheca macularis** (*S. humilis*); AK, CO, MT, TX, WY.

Rot, Root. **Phymatotrichum omnivorum**, TX.

Rot, White Heart. **Fomes fraxinophilus**, CO, MT, NM, SD, WY.

Rust. **Puccinia caricis-shepherdiae** (0, I), Rocky Mts., AK, MI, NY, OR, WA; II, III on *Carex*; **P. coronata** (0, I), SD to NM, AK, WA.

BUMELIA

Leaf Spot. **Cercospora lanuginosa**, TX; **Phyllosticta bumeliifolia**, AL, TX; **P. curtisii**, FL, MO; **Septoria bumeliae**, MS.

Rot, Root. **Helicobasidium purpureum**, TX; **Phymatotrichum omnivorum**, TX.

BUNDLEFLOWER (*Desmanthus*)

Leaf Spot. **Cercospora desmanthi**, KS, LA, MO, NM, SD, TN.

Powdery Mildew. **Erysiphe polygoni**, IL, MS.

Rust. **Ravenelia texensis** (II, III), TX.

BUTTERFLY PEA (*Centrosema*)

Leaf Spot. **Cercospora clitoriae**, AL; **Colletotrichum** sp., on pods.

BUTTERFLY PEA (*Clitoria*)

Leaf Spot. **Cercospora clitoriae**, AL, FL; **C. cruenta**, FL.

Rot, Root. **Phymatotrichum omnivorum**, TX.

BUTTONBUSH (*Cephalanthus*)

Blight, Leaf. **Cercospora perniciosa**, TX.

Blight, Thread. **Pellicularia koleroga**, FL.

Dodder. **Cuscuta compacta**, FL; **C. gronovii**, NY.

Leaf Spot. **Ascochyta cephalanthi**, LA: **Coniothyrium cephalanthi**, LA; **Phyllosticta cephalanthi**, TX; **Ramularia cephalanthi**, AL, KS, LA, NY, WI; **Septoria cephalanthi**, KS, WI.

Powdery Mildew. **Microsphaera alni**, widespread. **Phyllactinia corylea**, IN.

Rust. **Puccinia seymouriana** (0, I), New England to FL and central states; II, III on *Spartina*, **Uredo cephalanthi** (II), FL.

C

CAESALPINIA
ANTHRACNOSE. **Gloeosporium** sp., FL.
BACTERIAL Crown Gall. **Agrobacterium tumefaciens**, FL.
CANKER; DIEBACK. **Botryosphaeria ribis** var. **chromogena**, FL, TX.
ROT, Root. **Clitocybe tabescens**, FL; **Phymatotrichum omnivorum**, TX.
RUST. **Ravenelia humphreyana** (II, III), FL.

CALIFORNIA-LAUREL (*Umbellularia*)
ANTHRACNOSE. **Kabatiella phorodendri** f. sp. **umbellulariae**, CA.
BACTERIAL Leaf Spot. **Pseudomonas lauraceum**, CA.
BLACK MILDEW. **Asterina anomala**, CA.
BLIGHT Leaf, **Colletotrichum gloeosporioides**.
CANKER; DIEBACK. **Nectria cinnabarina**; **N. coccinea**, CA.
LEAF SPOT, Sudden Oak Death. **Phytophthora ramorum**, CA.
NEMATODE, Pin. **Paratylenchus (Gracilacus) anceps**. Ring, **Criconemoides xenoplax**, CA; Sheath **Hemicycliophora brevis**, CA.
NEMATODE, Root Knot. **Meloidogyne hapla**, CA.
ROT, Wood. **Fomes** spp., **Lenzites betulina**; **Polyporus versicolor**; **Porioambigua**; **P. ferruginosa**; **Stereum albobadium**; **Schizophyllum commune**.

CALIFORNIA PEPPER-TREE (*Schinus*)
DODDER. **Cuscuta subinclusa**, CA.
LEAF SPOT. **Rhizoctonia solani**, FL.
NEMATODE, Root Knot. **Meloidogyne** sp., TX.
ROT, Heart. **Fomes applanatus**, CA; **Polyporus dryophilus**, CA; **P. farlowii**, AZ, CA; **P. sulphureus**; **P. versicolor**, CA.
ROT, Root. **Armillaria mellea**, CA; **Phymatotrichum omnivorum**, TX.
ROT, Wood. **Ganoderma polychromum**, CA; **Stereum hirsutum**; **Trametes hispida**, CA; **Schizophyllum commune**.
WILT. **Verticillium albo-atrum**, CA; **Raffaelea lauricola**, CA.

R.K. Horst, *Field Manual of Diseases on Trees and Shrubs*,
DOI 10.1007/978-94-007-5980-0_6,

CALLIANDRA (False-Mesquite)
ROT, Root. **Clitocybe tabescens**, FL.
RUST. **Ravenelia reticulatae** (II, III), AZ.

CALLICARPA (Beauty-Berry; French-Mulberry)
BLACK MILDEW. **Meliola cookeana**, FL, LA.
LEAF SPOT. **Cercospora callicarpae**, SC to TX.
NEMATODE, Burrowing. **Radopholus similis**, FL.

CALYCANTHUS (Sweetshrub, Carolina Allspice)
BACTERIAL Crown Gall. **Agrobacterium tumefaciens**, MS, NY.
CANKER. **Botryosphaeria calycanthi**, NC, VA.
POWDERY MILDEW. **Phyllactinia corylea**, CA.

CAMELLIA
BACTERIAL Crown Gall. **Agrobacterium tumefaciens**, WA.
BLACK MILDEW. **Meliola camelliae**.
BLIGHT. **Phytophthora ramorum**, CA, OR, WA.
BLIGHT, Flower. **Sclerotinia camelliae**, CA, GA, LA, MS, NC, SC, TX, VA; **S. sclerotiorum**, NC; **Pestalotia** sp., **Penicillium** sp., AL.
BLIGHT, Gray Mold; Bud and Flower. **Botrytis cinerea**, general after frost.
BLIGHT, Petal. **Pestalotia** sp., AL.
CANKER; DIEBACK. **Glomerella cingulata**, widespread; **Botryosphaeria ribis**, MS; **Phomopsis** sp., FL.
CANKER, Felt Fungus. **Septobasidium castaneum**; **S. conidiophorum**; **S. pseudopedicellatum**, SC.
DODDER. **Cuscuta** sp.
GALL, Leaf. **Exobasidium camelliae**, and var. **gracilis** on *Camellia sasanqua*, FL, LA, MS, TX; **E. monosporum**, AL.
LEAF SPOT. **Cercospora** sp., GA; **C. theae**, LA; **Hendersonia subalbicans**, GA; **Pestalotia quepini**, also twig blight, widespread; **Phyllosticta camelliae** and **P. camelliaecola**, Southeast; **Sporonema camelliae**.
LEAF SPOT, Algal. **Cephaleuros virescens**, Gulf states.
NEMATODE, Burrowing. **Radopholus similis**, FL.
NEMATODE, Lesion. **Pratylenchus coffeae**.
NEMATODE, Ring. **Criconemoides komabaensis**, FL; **Hemicriconemoides gaddi**, GA, LA.
NEMATODE, Root Knot. **Meloidogyne incognita**.
NEMATODE, Spiral. **Helicotylenchus nannus**.
NONPARASITIC. **Bud Drop**, **Dieback**. Malnutrition, freezing, desiccation, widespread.
Oedema. Corky excrescences, from disturbed water relations.
Sunscald. Light circular spots, often with secondary fungi.
ROT, Root. **Phytophthora cinnamomi**, major problem, AL, CA; **Clitocybe tabescens**, FL, LA.
SPOT ANTHRACNOSE, Scab. **Elsinoë leucospila**, FL, GA, LA.

VIRUS. **Camellia Yellow Mottle Leaf** (color-breaking strains).

Flower blight is a devastating disease that has spread with plants in cans or pots. Order camellias bare-rooted, with all flower buds showing color picked off. Dieback is the subject of much controversy among camellia fans. Drastic surgery and a copper spray early in the season to prevent infection through bud scars seems to be helpful. Chelated iron may mask color-breaking virus symptoms.

CAMPHOR-TREE (*Cinnamomum*)
ANTHRACNOSE. **Glomerella cingulata**, Gulf states.
BLACK MILDEW. **Lembosia camphorae**, FL.
CANKER; DIEBACK. **Diplodia camphorae**, **D. natalensis**, **D. tubericola**, widespread; **Gloeosporium camphorae** and **G. ochraceum**, also leaf spot, Gulf states.
LEAF SPOT, Algal. **Cephaleuros virescens**, FL to LA.
MISTLETOE. **Phoradendron serotinum** (**flavescens**), FL.
NEMATODE, Lesion. **Pratylenchus pratensis**, FL.
NONPARASITIC. **Chlorosis**. Manganese deficiency, FL.
POWDERY MILDEW. **Microsphaera alni** var. **cinnamomi**, LA.
ROT, Root. **Armillaria mellea**, FL; **Phymatotrichum omnivorum**, TX; **Clitocybe tabescens**, FL.
SPOT ANTHRACNOSE. **Elsinoë cinnamomi**, MS, SC.
Wilt, Laurel. **Raffaelea lauricola**, GA, FL.

CANDLESTICK SHRUB (*Cassia*)
LEAF SPOT. **Tubakia dryina**, LA.

CAPE-HONEYSUCKLE (*Tecomaria*)
ANTHRACNOSE. **Colletotrichum gloeosporioides**, TX.
ROT, Root. **Armillaria mellea**, CA; **Clitocybe tabescens**, FL.

CAPER (*Capparis*)
BLACK MILDEW. **Asterina lepidigenoides**, FL; **A. radians**, FL.

CARISSA
CANKER; DIEBACK. **Physalospora obtusa** and **P. rhodina**, FL.
LEAF SPOT. **Colletotrichum gloeosporioides**, FL; **Macrophoma** sp., CA; **Septoria** sp., CA.
NEMATODE, Root Knot. **Meloidogyne** sp., CA, FL.
ROT, Root. **Phymatotrichum omnivorum**, TX.

CARISSA (Natal-Plum)
CANKERS; GALL. **Sphaeropsis tumefaciens**, FL.

CAROB, ST. JOHNS BREAD (*Ceratonia*)
CANKER. **Botryosphaeria ribis**, CA.
NEMATODE, Root Knot. **Meloidogyne** spp.
ROT, Root. **Phymatotrichum omnivorum**, TX; **Phytophthora cactorum**, CA.
WILT. **Verticillium albo-atrum**, CA.

CAROLINA JESSAMINE (*Gelsemium*)
BLACK MILDEW; Black Spot. **Asterina somatophora**, FL.
LEAF SPOT. **Phyllosticta gelsemii**, NJ.
ROT, Root. **Phymatotrichum omnivorum**, TX.
SOOTY MOLD. **Capnodium grandisporum**, MS, TX.

CAROLINA MOONSEED (*Cocculus*)
LEAF SPOT. **Cercospora cocculicola**, OK; **C. menispermi**, MS, TX.
NEMATODE, Burrowing. **Radopholus similis**, FL.
ROT, Root. **Phymatotrichum omnivorum**, TX.

CASHEW (*Anacardium*)
BLIGHT, Seedling. **Sclerotium rolfsii**, FL.

CASSIA (Senna, Sicklepod, Partridgepea)
BLIGHT, Seedling. **Alternaria cassiae**, MS.
CANKER, Stem. **Diaporthe phaseolorum** var. **caulivora**, LA.
DIEBACK. **Diplodia natalensis**, TX.
LEAF SPOT. **Cercospora nigricans**, MS; **Septoria cassiicola**; **Tubakia dryina**, LA.
NEMATODE, Root Knot. **Meloidogyne** sp., CA, TX.
POWDERY MILDEW. **Microsphaera alni**, MD; **Erysiphe polygoni**.
ROT, Black. **Cylindrocladium parasiticum**, GA.
ROT, Root. **Clitocybe tabescens**, FL; **Phymatotrichum omnivorum**, TX.
WILT. **Fusarium oxysporum** f. sp. **cassiae**, FL, GA, NC, SC.

CASSIOPE
LEAF GALL. **Exobasidium vaccinii**, WA.

CASUARINA (Australian-Pine)
NEMATODE, Root Knot. **Meloidogyne** sp., FL.
ROT. **Armillaria mellea**, CA; **Clitocybe tabescens**, FL. *Casuarina* is particularly susceptible to Clitocybe root rot but *C. cunninghamiana* is more resistant than other species.

CATALINA CHERRY (*Prunus lyonii*)
LEAF SPOT; Shot Hole. **Coryneum beierincki**.

CATALPA
ANTHRACNOSE. **Gloeosporium catalpae**, MD, MA, NJ, NY, PA.
BLIGHT, Southern. **Sclerotium rolfsii**, on seedlings.
DAMPING-OFF. **Rhizoctonia solani**, NE; **Pythium ultimum** (somewhat resistant).
LEAF SPOT. **Alternaria catalpae**, widespread; **Cercospora catalpae**, MA to FL, TX and IA; **Phyllosticta catalpae**, general.
NEMATODE, Root Knot. **Meloidogyne** sp., southern states to OH.
NONPARASITIC. **Chlorosis**. Soil alkalinity. **Leaf Scorch**. Heat, drought.
POWDERY MILDEW. **Microsphaera alni** var. **vaccinii**, MA to AL, IL, NE, TX; **Phyllactinia corylea**, IL, IN, KY, MA, OH, SC, VA.

Rot, Root. **Armillaria mellea**, WA; **Helicobasidium purpureum**, OH; **Phymatotrichum omnivorum**, TX; **Thielaviopsis basicola**, of seedlings.
Rot, Wood. **Collybia velutipes**, IN; **Polyporus** spp.; **Schizophyllum commune**; **Stereum** spp.; **Trametes sepium.**
Sooty Mold. **Capnodium axiliatum**, LA, SC.
Spot Anthracnose. **Sphaceloma** sp.
Wilt. **Verticillium albo-atrum**, IL, IN, KS, MA, NJ, NY, OH, VA.

Leaf spots may cause some defoliation in a wet season, but many years the expense of spraying may be unjustified on a limited budget. Verticillium wilt kills street trees.

CATHA (Arabian-Tea)
Blight, Leaf Tip, **Colletotrichum gloeosporioides**, FL.

CATS-CLAW (*Doxantha*)
Rot, Root. **Phymatotrichum omnivorum**, TX.

CEANOTHUS
Bacterial Crown Gall. **Agrobacterium tumefaciens**, WA.
Leaf Spot. **Cercospora ceanothi**, KS, WI; **Septoria ceanothi**, ID; **Cylindrosporium ceanothi**, Pacific Coast; **Phyllosticta ceanothi**, MS.
Powdery Mildew. **Microsphaera alni**, widespread.
Rot, Root. **Armillaria mellea**, CA.
Rot, Sapwood. **Schizophyllum commune**, CA.
Rust. **Puccinia tripsaci** (0, I), KS, NE, WI; II, III on grasses.
Wilt. **Verticillium albo-atrum**, CA.

CEDAR (*Cedrus* spp.) (Atlas Cedar, Deodar, Cedar of Lebanon)
Canker; Dieback. **Diplodia pinea**, AL; **Sirococcus conigenus**, OR.
Rot, Heart. **Fomes pini**, occasional.
Rot, Root, **Armillaria mellea**, MS; **Clitocybe tabescens**, FL; **Phymatotrichum omnivorum** TX; **Phytophthora cinnamomi**, CA.

CEPHALOTAXUS (Japanese Plum-Yew)
Blight, Nursery. **Phomopsis juniperovora**, NY.

CESTRUM
Bacterial Canker, Vascular. **Clavibacter michiganense**, WY.

CHAMAECYPARIS (Atlantic White-Cedar, Port Orford White-Cedar, Alaska Yellow-Cedar, Hinoki Cypress, Sawara Cypress-Retinospora)
Bacterial Crown Gall. **Agrobacterium tumefaciens**, CA.
Blight, Nursery. **Phomopsis juniperovora**, widespread.
Blight, Tip. **Pestalotia funerea**, MI, NJ, TX; **Didymascella chamaecy-parissi**, NY.
Canker, Bark Patch. **Aleurodiscus nivosus**, NJ.
Nematode. **Sphaeronema** sp., AK; **Crossonema** sp., AK.
Nonparasitic. **Scorch**. Sun, freezing, drought, mites.

Rot, Collar and Root. **Phytophthora lateralis**, CA, OR, WA, serious; **P. cinnamomi**.

Rot, Heart. **Fomes pini**, occasional; **F. pinicola**, AK; **F. subroseus**, NJ, NC.

Rot, Root. **Armillaria mellea**, VA; **Clitocybe tabescens**, FL; **Phymatotrichum omnivorum**, TX; **Pythium ultimum**, CA; **Phytophthora lateralis**, CA.

Rot, Wood. **Lenzites saepiaria**, VA; **Poria** spp.; **Polyporus** spp.; **Steccherinum balloui**, NJ; **Trametes isabellina**, CA; **Fomes annosus**.

Rust, Gall. **Gymnosporangium biseptatum** (III), ME, NH to NJ, AL; 0, I on service-berry; **G. nootkatense** (III), AK, OR, WA; 0, I on crabapple, pear, mountain-ash; **G. fraternum** (III), ME, MA, NJ; 0, I on chokeberry.

Rust, Witches' Broom. **Gymnosporangium ellisii** (III), ME to FL and AL; 0, I on sweet fern, bayberry, wax-myrtle.

CHAMAEDAPHNE (Cassandra, Leatherleaf)

Gall, Leaf. **Exobasidium vaccinii**, widespread; **Synchytrium vaccinii**, NJ.

Leaf Spot. **Ascochyta cassandrae**, NY to WI; **Venturia arctostaphyli**, NH, NY.

Rust. **Chrysomyxa ledi** var. **cassandrae** (II, III), ME to PA and MN; AK; 0, I on pine.

CHERIMOYA, CUSTARD-APPLE (Annona)

Anthracnose, Fruit Rot. **Glomerella cingulata**, FL.

Blight, Stem. **Diplodia natalensis**, TX.

Rot, Root. **Armillaria mellea**, CA; **Clitocybe tabescens**, FL; **Phymatotrichum omnivorum**, TX.

Rust. **Phakopsora cherimoliae** (II, III), FL, TX.

CHERRY (*Prunus* spp.)

Bacterial Black Spot; Canker; Gummosis. **Xanthomonas pruni**, NY to MI, GA and TX.

Bacterial Canker. **Pseudomonas syringae**, MA to MI; Pacific states and **P. morsprunorum**.

Bacterial Crown Gall. **Agrobacterium tumefaciens**, CA, TX, WA.

Bacterial Fire Blight. **Erwinia amylovora**, OR, WA.

Bacterial, MLO. **Peach X-Disease**; **Peach Yellow Leaf Roll** (**Peach Western X-Disease**), Northwest.

Black Knot. **Dibotryon morbosum**, eastern states.

Blight, Blossom, Brown Rot. **Monilinia laxa**, general, Pacific states, MI, NY, WI; **Sclerotinia sclerotiorum**, OR.

Blight, Seedling, Twig. **Monilinia rhododendri**, VT to GA, AR, IA; **M. fructicola**, shoot and leaf, GA.

Blight, Shoot; Shot Hole. **Coryneum carpophilum** (*C. beijerinckii*), CA, ID, OR, WA; **Monilinia laxa**; MA, NY, RI.

Canker, Felt Fungus. **Septobasidium retiforme**, GA.

Canker, Trunk and Collar. **Phytophthora cactorum**, **P. citrophthora**, CA.

Canker, Twig. **Phomopsis padina**; **Cytospora leucostoma**, NY, PA; **Valsa leucostoma**, widespread; **Nectria** sp., NY; **Eutypa lata** (syn. **E. armeniacae**), CA.

Dieback, Branch. **Calosphaeria pulchella**, CA.
Leaf Blister. **Taphrina farlowii**, VT to FL, TX; **Taphrina cerasi**, ME to NJ, MN.
Leaf Spot. **Cercospora circumscissa** (*Mycosphaerella cerasella*), NJ, PA, VA to FL, TX; **Alternaria citri** var. **cerasi**, CA; **Phyllosticta pruni-avium**, OR; **Cercospora graphioides** and **Phyllosticta serotina** on wild black cherry only; **Coccomyces lutescens**, VA.
Leaf Spot; Blight; Shot hole. **Coccomyces hiemalis**, general.
Mistletoe. **Phoradendron serotinum (flavescens)**.
Nematode, Lesion. **Pratylenchus** sp.; **P. vulnus**, CA.
Nematode, Root Knot. **Meloidogyne** sp.; **M. incognita**; **M. javanica**.
Nonparasitic. **Brown Bark Spot**. Arsenical poisoning of soil, ID, MT, WA. **Chlorosis**. Alkali injury, CA, TX. **Little Leaf**. Zinc deficiency, CA, OR, WA.
Powdery Mildew. **Podosphaera oxyacanthae**, general.
Rot, Brown. **Monilinia fructicola**, general; **M. laxa**, MA, NY, RI, Argentina.
Rot, Fruit. **Alternaria** sp.; **Botrytis cinerea; Cladosporium herbarum**, CA to WA and ID; **Lambertella** sp., OR; **Microstroma tonellianum**, MA; **Penicillium expansum; Rhizopus stolonifer**, occasional in market; **Pullularia** sp., Northwest; **Mucor piriformis.**
Rot, Green Fruit. **Sclerotinia sclerotiorum**, OR.
Rot, Heart; Wood. **Fomes fomentarius**, Northeast; **Polyporus** spp.; **Poria ambigua**, CA.
Rot, Root. **Armillaria mellea**, NM, OK, OR; **Phymatotrichum omnivorum**, TX; **Xylaria** spp., VA; **Poria ambigua**, CA.
Rot, Silver Leaf. **Stereum purpureum**, NY, MT.
Rust. **Tranzschelia discolor** (II, III), GA, MA, NE, NY, NC, OK, TX; 0, I on anemone.
Scab. **Fusicladium cerasi**, NY to IA and WI.
Virus. **Cherry Rough Fruit**, UT on sweet cherry (*Prunus avium*); **Cherry Albino**; **Cherry Bark-Splitting**; **Cherry Black Canker; Cherry Buckskin**; **Cherry Bud Abortion**; **Cherry Chlorosis**; **Cherry Freckle Fruit Disease**; **Cherry Green Ring Mottle**; **Cherry Gummosis**; **Cherry Little Cherry**; **Cherry Midleaf Necrosis**; **Cherry Mora**; **Cherry Mottle Leaf**; **Cherry Necrotic Rusty Mottle**; **Cherry Pink Fruit**; **Cherry Pinto Leaf**; **Cherry Rasp Leaf**; **Cherry Ring Spot**; **Cherry Rugose Mosaic**; **Cherry Rusty Mottle**; **Cherry Twisted Leaf**; **Cherry Vein Clearing**; **Cherry Yellows**; **Prune Dwarf**; **Tomato Ring Spot**; **Stem Pitting**; **Prunus Necrotic Ring Spot**, CA, WA; **Plum Pox**, WA.
Virus, Leafroll and Yellow Leaves. **Nepovirus**, WA.

CHERRY, FLOWERING, ORIENTAL (*Prunus serrulata*)
Bacterial Fire Blight. **Erwinia amylovora**, GA, OH.
Bacterial Leaf Spot. **Xanthomonas pruni**, NJ.
Canker; Dieback. **Botryosphaeria ribis**, GA.
Decline. Cause, unknown; complex of insect, poor soil, viral infection and air pollutants suspected, VA.
Leaf Blister; Witches' Broom. **Taphrina cerasi**, MD, NJ.

LEAF SPOT. **Coccomyces hiemalis**, MA, NJ.

SCAB. **Cladosporium carpophilum**, MS.

VIRUS. **Cherry Vein Clearing**; **Cherry (Flowering) Rough Bark**; **Cherry Little Cherry**.

CHERRY, JAPANESE FLOWERING (*Prunus subhirtella*)

LEAF SPOT. **Cristulariella pyramidalis**, FL.

CHERRY-LAUREL (*Prunus laurocerasus*)

BACTERIAL Spot. **Xanthomonas pruni**, GA, MS, NJ, SC.

BLIGHT, Blossom; Brown Rot. **Monilinia fructicola**, CA; **M. laxa**, CA; **Alternaria** sp., TX.

LEAF SPOT. **Cercospora circumscissa**, CA; **C. cladosporioides**, LA, TX; **Coccomyces lutescens**, MS; **Phyllachora beaumontii**, AL; **Phyllosticta laurocerasi**, CA, FL, NJ; **Septoria ravenelii**, SC.

MISTLETOE. **Phoradendron serotinum (flavescens)**, FL.

ROT, Root. **Armillaria mellea**, CA; **Clitocybe tabescens**, FL; **Phymatotrichum omnivorum**, TX.

WILT. **Verticillium albo-atrum**, CA.

CHESTNUT (*Castanea*)

BLIGHT, Canker. **Endothia parasitica**, general, with American chestnut practically exterminated by it.

BLIGHT, Twig. **Cytospora** sp.; **Phomopsis** sp.; **Diplodia longispora**.

CANKER, Bark Patch. **Aleurodiscus aceris**.

CANKER; DIEBACK. **Cryptodiaporthe castanea**; **Strumella coryneoidea**.

LEAF SPOT. **Actinopelte dryina**; **Cylindrosporium castaneae**; **Marssonina ochroleuca**, general; **Monochaetia desmazierii**; **M. kansensis**; **Phyllosticta castanea**; **Exosporium fawcettii**; **Scolecosporium fagi**.

MISTLETOE. **Phoradendron villosum**, CA.

POWDERY MILDEW. **Microsphaera alni**; **Phyllactinia corylea**.

ROT, Root. **Armillaria mellea**; **Phytophthora cinnamomi**; **Phymatotrichum omnivorum**.

ROT, Wood. **Fomes** spp.; **Polyporus** spp.; **Poria** spp.; **Stereum** spp.

Japanese and Chinese chestnuts are resistant to chestnut blight. Plant breeders are trying to develop hybrids between Asiatic and native species that will be resistant to *Endothia*.

CHINABERRY (*Melia*)

BLACK MILDEW. **Meliola** sp.

BLACKLEG. **Phoma lingam**, WA.

BLIGHT, Limb; Twig. **Pellicularia koleroga**, FL; **Eutypella stellulata**, OK, TX; **Fusarium lateritium**, TX.

CANKER. **Nectria coccinea**, MS, SC.

DOWNY MILDEW. **Peronospora parasitica**, general.

LEAF SPOT. **Cercospora leucosticta**, Gulf states; **C. meliae**; **C. subsessilis**; **Phyllosticta azedarachis**, AL; **P. meliae**, LA, TX.

MISTLETOE. **Phoradendron serotinum (flavescens)**, TX.
NEMATODE, Root Knot. **Meloidogyne** sp.
PHYTOPLASMA; Witches' Broom and Decline, Brazil, Rio Grande do Sul.
POWDERY MILDEW. **Phyllactinia corylea**, MS.
ROT, Root. **Helicobasidium purpureum**, TX; **Phymatotrichum omnivorum**, AZ, TX.
ROT, Wood. **Fomes meliae**, AL; **Polyporus versicolor**, GA.

CHINESE TALLOWTREE (*Sapium*)
LEAF SPOT. **Cercospora stillingiae**, LA; **Phyllosticta stillingiae**, LA.
ROT, Root. **Clitocybe tabescens**, FL; **Phymatotrichum omnivorum**, TX.

CHINQUAPIN (*Castanopsis*)
BLIGHT, Chestnut; Canker. **Endothia parasitica**, general.
CANKER, Brown Felt. **Septobasidium pseudopedicellatum**.
LEAF BLISTER. **Taphrina castanopsidis**, CA.
LEAF SPOT. **Dothidella castanopsidis**, CA, OR; **Mycosphaerella** sp., CA, OR.
POWDERY MILDEW. **Microsphaera alni**, NC.
ROT, Heart; Sapwood; Wood. **Fomes igniarius**, OR; **Ganoderma oregonenses**; **Peniophora sanguinea**; **Polyporus hirsutus**; **P. versicolor**.
ROT, Root. **Armillaria mellea**, OR.
WILT, Oak. **Ceratocystis fagacearum**.

CHIOGENES (*Creeping Snowberry*)
RUST. **Chrysomyxa chiogenis** (II, III), MI, NH, NY, WI; 0, I unknown.

CHOKEBERRY (*Aronia*)
BACTERIAL Fire Blight. **Erwinia amylovora**, MI, TX, WV.
DODDER. **Cuscuta compacta**, FL.
LEAF SPOT. **Ascochyta pirina**, WI; **Cercospora mali**, AL, TX; **C. piri**, MI, NH, WI; **Mycosphaerella arbutifolia**, NY; **Phyllosticta arbutifolia**, NJ.
ROT, Brown. **Monilinia fructicola**, WI.
ROT, Root. **Phymatotrichum omnivorum**, TX.
RUST. **Gymnosporangium clavariiforme** (0, I), MA; III on juniper; **G. clavipes** (0, I), CT, ME, MA, TX; **G. davisii** (0, I), ME, MI, NH, WI; **G. fraternum** (0, I), DE, MA, ME, NJ, PA; III on chamaecyparis.

CHOKECHERRY (*Prunus virginiana*)
BACTERIAL Leaf Spot. **Xanthomonas pruni**, IL, MT, NY, WY.
BACTERIAL, MLO. **Peach X-Disease**, NH to VA, IL, WI.
BLACK KNOT. **Dibotryon morbosum**, general.
BLIGHT, Fruit and Shoot. **Monilinia demissa**, ID, WA; **M. padi** (*Sclerotinia angustior*), VT to KS, ND.
CANKER; DIEBACK. **Cytospora chrysosperma**, MT.
CANKER, TWIG. **Eutypa lata** (syn. **E. armeniacae**), CA.
LEAF BLISTER; Fruit, Shoot Hypertrophy. **Taphrina confusa**, widespread.
LEAF SPOT. **Cylindrosporium nuttallii**, OR; **Gloeosporium prunicola**, NY; **Lophodermina prunicola**, tar spot, CO; **Mycosphaerella cerasella**, KS;

Phyllosticta circumscissa; **P. virginiana**, leaf blotch, NY to KS and MT; **Septoria pruni**, MI.
LEAF SPOT; Shot Hole. **Cercospora circumscissa**, New England, IA, WI; ND to KS, and MT; **Cercospora lutescens**, general.
NEMATODE, Root Knot. **Meloidogyne** sp.
POWDERY MILDEW. **Podosphaera oxyacanthae**, widespread; **Phyllactinia corylea**, WA.
ROT, Brown Heart. **Fomes fulvus**, ND, SD.
RUST. **Tranzschelia pruni-spinosae** (II, III), CT to IL, WI.
VIRUS. **Western X-Disease**, ID, OR, UT, WA; **Tatter Leaf** (**Peach Ring Spot**).

Chokecherries should be eliminated near peach orchards to control the peach X-disease.

CHRISTMAS BERRY (*Heteromeles arbutifolia*)
BLIGHT, Sudden Oak Death. **Phytophthora ramorum**, CA.
RUST. **Gymnosporium exiguum**, CA.

CINCHONA
LEAF SPOT. **Cercospora cinchonae**, LA.
NEMATODE, Root Knot. **Meloidogyne** sp.

CINNAMON-TREE (*Cinnamomum zeylandicum*)
ANTHRACNOSE. **Glomerella cingulata**, FL.
LEAF SPOT, Algal. **Cephaleuros virescens**, FL.

CISSUS
LEAF SPOT. **Cercospora viticola**, LA; **C. arboreae**, TX; **Phyllosticta cissicola**, TX.
ROT, Root. **Phymatotrichum omnivorum**, TX.
RUST. **Aecidium mexicanum** (0, I), OK.
SMUT. **Mycosyrinx cissi**, FL.

CITRUS FRUITS (Grapefruit, Lemon, Lime, Orange)
ANTHRACNOSE; Withertip, Fruit Spot. **Glomerella cingulata**, general.
ANTHRACNOSE, Lime; Withertip. **Gloeosporium limetticolum**, on lime only, CA.
ANTHRACNOSE, Scab. **Elsinoë australis**, FL.
BACTERIAL Blast; Blight; Black Pit. **Pseudomonas syringae**, CA; after cold rains.
BACTERIAL Brown Leaf Spot. **Burkholderia andropogonis**, FL.
BACTERIAL Canker. **Xanthomonas campestris** pv. **citri**, FL.
BACTERIAL Crown Gall. **Agrobacterium tumefaciens**, AZ, CA.
BACTERIAL Greening. **Candidatus Liberibacter asiaticus**, Central America, Puerto Rico.
BACTERIAL, MLO. **Citrus Stubborn Disease**.
BLIGHT, Brown Spot; Fruit and Young Shoots. **Alternaria citri**, FL.
BLIGHT, Leaf and Stem. **Phytophthora syringae**, CA.
BLIGHT, Seedling; Fruit Rot. **Sclerotium rolfsii**.
BLIGHT, Thread. **Pellicularia koleroga**, Gulf states.

BLIGHT, Twig. **Gibberella baccata** (*Fusarium lateritium*), CA; **Tryblidiella rufula**, TX.

BLOTCH, Sooty. **Gloeodes pomigena**, Gulf states.

CANKER, Branch. **Eutypella** spp., CA; **Spencermartinsia viticola**, **Neofusicoccum austral**, and **N. parvum**, CA.

CANKER, Branch Knot. **Sphaeropsis tumefaciens**, FL.

CANKER, Branch Wilt. **Exosporina fawcettii**, CA.

CANKER; DIEBACK. **Aspergillus foetidus**, CA.

CANKER, Felt Fungus. **Septobasidium** spp., Gulf states.

CANKER, Trunk. **Botryodiplodia theobromea**, TX.

CANKER, Wound. **Hendersonula toruloidea**, CA.

DAMPING-OFF. **Pythium** spp.; **Rhizoctonia solani**, cosmopolitan.

DODDER. **Cuscuta americana**, CA, FL; **Cassytha filiformis**, a dodderlike plant, FL.

GREENING DISEASE. **Huanglongbing**, FL.

LEAF SPOT. **Cercospora aurantia**, FL, TX; Greasy Spot, **C. citri-grisea**, FL; Tar Spot, **C. gigantea**, FL; **Cladosporium oxysporum**, FL; **Mycosphaerella lageniformis**, CA; **M. citri**, FL; **Pleospora** sp., CA; **Alternaria citri**, on Rangpur lime; Fruit Spot, **Septoria citri**, CA, TX; **S. limonium**, CA; **Xanthomonas fuscans**, Brazil.

LEAF SPOT, Algal; Red Rust. **Cephaleuros virescens**, Gulf states.

MISTLETOE. **Phoradendron** sp., Gulf states.

NEMATODE, Burrowing. **Radopholus similis**, FL.

NEMATODE, Citrus. **Tylenchus semipenetrans**, CA, FL.

NEMATODE, Lance. **Hoplolaimus coronatus**.

NEMATODE, Lesion. **Pratylenchus brachyurus**; **P. vulnus**.

NEMATODE, Ring. **Criconema civellae**; **Criconemoides citri**, FL.

NEMATODE, Stubby Root. **Paratrichodorus christiei**.

NONPARASITIC. **Blight**. Boron deficiency, CA, FL.

Bronzing. Magnesium deficiency, FL. **Cancroid Spot.** Genetic abnormality.

Chlorosis. Iron deficiency, AZ, CA, FL; manganese deficiency, FL.

Exanthema. Copper deficiency, augmented by excessive nitrogen fertilization and bad drainage, CA, FL.

Greasy Spot. Black Melanose, Cause unknown, CA, TX; chiefly grapefruit.

Gummosis. Gum Spot; Gummosis. Environmental injuries in part.

Leprosis. Florida Scaly Bark; Nailhead Rust. Cause unknown, FL.

Mottle Leaf. Zinc deficiency, AZ, CA, FL.

Oleocellosis. Rind-oil Spot. Chemical injury from release of oil in rind.

Rumple. Rind network on lemon; cause unknown.

Silver Scurf. Thrips injury on fruit, widespread; Silvering. Rust mite injury; Stigmonose. Insect punctures.

Wilt; Blight; Leaf Curl. Irregular water supply, FL.

PHYTOPLASMA, Citrus Huanglongbing. **Candidatus L. asiaticus**, FL.

POWDERY MILDEW. **Oidium** sp., CA, FL; **O. tingitaninum**.

ROT, Black, of Fruit. **Alternaria citri**, widespread.

Rot, Brown; Gummosis. **Phytophthora citrophthora**, AZ, CA, FL; **P. parasitica**, "mal di gomma," CA, FL.
Rot, Charcoal. **Macrophomina phaseoli**, AZ, CA.
Rot, Cottony Fruit; Twig Blight. **Sclerotinia sclerotiorum**, CA, FL, TX.
Rot, Dothiorella; Bark Canker. **Botryosphaeria ribis**, widespread.
Rot. **Diplodia**; Collar; Twig Blight. **Diplodia natalensis**, general.
Rot, Fruit. **Aspergillus** spp.; **Fusarium** spp.; **Mucor** spp.; **Oospora citri aurantii**, sour rot, **Nematospora coryli**, dry rot; **Candida krusei**; **Gloeosporium foliicolum** (*Glomerella cingulata*), FL; **Penicillium italicum**, Mexico.
Rot, Gray Mold; Twig Blight. **Botrytis cinerea**, CA.
Rot, Green Mold. **Penicillium digitatum**; Blue Mold, CA; **P. italicum**; Pink Mold, **P. roseum**, cosmopolitan; Black Mold, **Rhizopus stolonifer**; **Trichothecium viride**, occasional.
Rot. Melanose; Phomopsis; Decorticosis; Shell Bark. **Diaporthe citri**, general.
Rot, Mushroom Root. **Armillaria mellea**, CA; **Clitocybe tabescens**, FL.
Rot, Root. **Fusarium solani**; **Phymatotrichum omnivorum**, TX; **Poriavaporaria**, CA; **Thielaviopsis basicola**.
Rot, Sour. **Geotrichum citri-aurantii**, Mexico.
Rot, Wood. **Daldinia concentrica**, occasional; **Ganoderma lucidum**, FL; **Polyporus** spp.; **Schizophyllum commune**; **Trametes hydnoides**; **T. hispida**.
Sooty Mold, **Capnodium citri**; **C. citricola**. Gulf states.
Spot Anthracnose; Citrus Scab. **Elsinoë fawcettii**, AL, CA, FL, LA, MS.
Virus. **Algerian Navel Orange**, FL; **Citrus Ringspot**, CA, FL; **Citrus Exocortis** (viroid); **Citrus Psorosis**, Scaly Bark; **Citrus Tatter Leaf**; **Citrus Tristeza**, Quick Decline; **Citrus Vein Enation**; **Citrus Xyloporosis**, Cachexia; **Citrus Yellow Vein**; **Citrus Leaf Rugose**, CA; **Citrus Tristeza**, LA; **Citrus Variegation**; **Citrus Leprosis**; **Citrus Enation**-Woody Gall, CA.

In the home garden oil sprays for scale help to get rid of sooty mold. A neutral copper spray just before growth starts is recommended in Florida for citrus scab and one just after flowering for melanose. Citrus trees in the burrowing nematode area are being pulled and the soil treated. Consult the California and Florida Agricultural Experiment Stations for the latest information on specific Citrus problems.

CLAUSENA (Wampi)
Leaf Spot. **Glomerella cingulata**, MD.
Nematode, Citrus. **Tylenchulus semipenetrans**, FL.
Spot Anthracnose; Citrus Scab. **Elsinoë fawcettii**, FL.

CLEMATIS (including Virgins-Bower)
Bacterial Crown Gall. **Agrobacterium tumefaciens**, MN, TX.
Blight, Leaf. **Phloeospora adusta**, TX.
Leaf Spot. **Ascochyta clematidina**, widespread; **Cercospora rubigo**,CA, WA; **C. squalidula**, widespread; **Cylindrosporium clematidis**, East and South; **Glomerella cingulata**, FL; **Phyllosticta clematidis**, MT, VA; **Ramularia**

clematidis, MT; **Septoria clematidis**, WA, WI; **Sphaerella applanata**, MT, TX.

NEMATODE, Root Knot. **Meloidogyne hapla**, ND.

POWDERY MILDEW. **Erysiphe polygoni**, widespread.

RUST. **Puccinia recondita** (0, I), Rocky Mts. and Pacific Coast; II, III on grasses; **P. pulsatillae** (III), CA; **P. stromatica** (III), AL; **Tranzschelia viornae** (II, III), TX.

SMUT. **Urocystis carcinodes**, UT.

CLERODENDRUM (Glorybower)

CANKER, Stem. **Kutilakesa pironii**, FL.

LEAF SPOT. **Septoria phylctaenioides**, SC; **Cercospora apii** f. sp. **clerodendri**, FL.

NEMATODE, Root Knot. **Meloidogyne** sp., MD.

VIRUS. **Zonate Ring Spot**, FL.

CLETHRA (Sweet Pepperbush, White-Alder)

GALL, Red Leaf. **Synchytrium vaccinii**, NJ.

LEAF SPOT. **Phyllosticta clethricola**, MD, NJ, TX.

ROT, Root. **Corticium galactinum**.

COCCOLOBA (Sea-Grape, Dove-Plum)

BLACK MILDEW. **Lembosia coccolobae**; **L. philodendri**; **L. portoricense**; **L. tenella**, FL.

LEAF SPOT. **Pestalotia coccolobae**, FL.

RUST. **Puccinia canaliculata**, GA; **Uredo coccolobae**; **U. uviferae** (II), FL.

COCOA (*Theobroma cacas*)

GALL. **Agrobacterium tumefaciens**, FL.

ROT, Pod. **Phytophthora palmivora**, Puerto Rico.

COCOA-PLUM (*Chrysobalanus*)

LEAF SPOT. **Cercospora chrysobalani**, FL.

LEAF SPOT, Algal. **Cephaleuros virescens**, FL.

COFFEE-BERRY (*Rhamnus californicus*)

BLIGHT, Sudden Oak Death. **Phytophthora ramorum**, CA.

LEAF SPOT. **Phaeosphaerella rhamni**, CA.

RUST. **Puccinia mesneriana** (III), CA.

SOOTY MOLD. **Capnodium** sp., CA.

COLUMNEA

VIRUS. **Tomato Spotted Wilt**, PA.

CORALBERRY (*Symphoricarpos orbiculatus*)

CANKER; Stem Gall. **Phomopsis** sp., MD.

LEAF SPOT. **Cercospora symphoricarpi**, KS, NE, TX.

POWDERY MILDEW. **Microsphaera diffusa**, general.

ROT, Berry. **Alternaria** sp., CT.

ROT, Root. **Helicobasidium purpureum**, TX; **Phymatotrichum omnivorum**, TX.

RUST. **Puccinia crandallii** (0, I), KS, MO, OK.
SPOT ANTHRACNOSE. **Sphaceloma symporicarpii**.

CORDIA
POWDERY MILDEW. **Golovinomyces cichoracearum**, TX.
ROT, Root. **Clitocybe tabescens**, FL; **Phymatotrichum omnivorum**, TX.

COTONEASTER
BACTERIAL Fire Blight. **Erwinia amylovora**, widespread.
BACTERIAL Hairy Root. **Agrobacterium rhizogenes**, central states.
CANKER, Twig Blight. **Physalospora obtusa**, NY to OH, TX; **Diplodia** sp., TX; **Gibberella baccata**, CA.
LEAF SPOT. **Fabraea maculata**, CA, IA; **Phyllosticta cotoneastri**, MD; **P. cydoniae**, MS.
POWDERY MILDEW. **Podosphaera** sp., CA.
ROT, Root. **Armillaria mellea**, CA; **Clitocybe tabescens**, FL; **Phymatotrichum omnivorum**, AZ, TX; **Phytophthora** sp.
SCAB. **Venturia** sp., WA.

COURSETIA
RUST. **Phragmopyxis acuminata** (0, III), AZ, CA.

COWANIA
RUST. **Phragmidium andersonii** (I, II, III), ID.

CRABAPPLE, FLOWERING (*Malus*)
ANTHRACNOSE. Northwestern. **Neofabraea malicorticis**, OR.
BACTERIAL Fire Blight. **Erwinia amylovora**, widespread.
BLIGHT, Thread. **Pellicularia koleroga**, LA.
BLOTCH, Sooty. **Gloeodes pomigena**, IN.
CANKER, Blister. **Nummularia discreta**, WV.
CANKER, Coral Spot; Twig Blight. **Nectria cinnabarina**, AK.
CANKER; DIEBACK. **Valsa leucostoma**, WV; **Physalospora obtusa**, MI; **Sphaeropsis pyriputrescens**, WA; **Phacidiopycnis washingtonensis** (also Fruit Rot), WA.
LEAF SPOT. **Cercosporella pirina**, IL, WI; **Coniothyrium pirinum**, WI; **Fabraea maculata**, AK, MD; **Illosporium malifoliorum**, WV; **Marssonina coronaria**, IN, IA, MO, WI; **Phyllosticta** sp., IA; **P. solitaria**, IN, IA, KS, OH, TX, WV; **P. zonata**, IA; **Septoria pyri**.
MISTLETOE, European. **Viscum album**, CA.
POWDERY MILDEW. **Oidium pyrinum**, WI; **Phyllactinia corylea**, WA; **Podosphaera leucotricha**, IL, IA; **P. oxyacanthae**, WA, WI.
ROT, Black, Fruit, Leaf Spot. **Physalospora obtusa**, eastern and central states.
ROT, Heart. **Ganoderma applanatum**.
ROT, Root. **Armillaria mellea**, WA; **Phymatotrichum omnivorum**, TX.
RUST. **Gymnosporangium clavipes**, CT, NJ; **G. globosum**, AL, IN, KS, NJ, SC, VA; **G. juniperi-virginianae**, cedar-apple rust, general; **G. libocedri**, AK, OR; **G. nelsonii**, AK, WA; **G. nootkatense**, AK; **G. yamadae**, DE, PA.
SCAB. **Venturia inaequalis**, general.

SCAB, Twig Infections. **Venturia inaequalis**, MA, OH, PA, RI.

Cedar-apple rusts are common and injurious on most native crabapples; asiatic varieties are usually resistant. Remove red-cedars in the vicinity.

CRANBERRY (*Vaccinium*, subgenus *Oxycoccus*)
DIEBACK. **Diaporthe vaccinii**, WI.
GALL, Red Leaf. **Synchytrium vaccinii**, NJ.
GALL; Shoot Hypertrophy; Rose-bloom. **Exobasidium vaccinii**, general; **E. oxycocci**, MA.
LEAF SPOT. **Discohainesia oenotherae**, also storage rot, MA, NJ, OR, WA, WV; **Mycosphaerella nigromaculans**, black spot, OR, WA; **Phyllosticta putrefaciens**, also berry rot, MA, NJ; **Ramularia multiplex**, MI, NY, WI; **Venturia compacta**, leaf smudge, general; **Cladosporium oxycocci**, leaf mold, NJ, WA.
NEMATODE. **Atylenchus decalineatus**.
NEMATODE, Sheath. **Hemicycliophora similis**.
NEMATODE, Stubby Root. **Paratrichodorus christiei**.
POWDERY MILDEW. **Microsphaera alni** var. **vaccinii**, AL, NJ, OH.
ROT, Berry; Blotch. **Acanthorhynchus vaccinii**, ME to NC, WI, OR, WA; **Ceuthospora lunata**, black rot, MA, NJ, OR, WA, WI; **Curvularia inaequalis**, IL, NJ, WI; **Sphaeronema pomorum**, NJ; **Sporonema oxycocci**, AK, ME, MA, NJ, OR, WA, WI.
ROT, Berry Speckle. Several fungi associated.
ROT, Bitter; Leaf Spot. **Glomerella cingulata** var. **vaccinii**, general; **Phyllosticta vaccinii**, WI.
ROT, Early; Scald, Blast. **Guignardia vaccinii**, general; **Phyllosticta vaccinii**, WI.
ROT, End. **Godronia cassandrae**, general.
ROT, Fairy Ring; Root. **Psilocybe agrariella** var. **vaccinii**, MA, NJ.
ROT, Fruit. **Acanthorhyncus vaccinii** and **A. alba**, North America; **Gliocephalotrichum bulbilium**, MA, NJ.
ROT, Hard; Twig Blight. **Monilinia oxycocci**, MA, ME, OR, WA, WI.
ROT, Root; DIEBACK. **Phytophthora cinnamomi**, MA.
ROT, Storage. **Penicillium** spp., cosmopolitan; **Diaporthe vaccinii**, general; **Botrytis cinerea**, gray mold, also blossom blight; **Alternaria** sp.; **Dematium** spp.; **Melanospora destruens**, in market; **Gloeosporium minus**, MD, NJ; **Pestalotia vaccinii**; **Stemphylium ilicis**.
ROT, Witches' Broom. **Naevia oxycocci**, ME, MI, NH, NY.
RUST. **Pucciniastrum vaccinii** (II, III), OR, UT, WA; 0, I on hemlock.
VIRUS. **Cranberry False-Blossom**, ME to NJ, OR, WA, WI; **Cranberry Ring Spot**, WI.

CRAPE-MYRTLE (*Lagerstroemia*)
BLIGHT, Thread. **Rhizoctonia ramicola**, FL.
LEAF SPOT. **Cercospora lythracearum**, TX; **Cercospora** sp., blotch, FL, TX; **Phyllosticta lagerstroemiae**, tip blight, LA, TX.
NONPARASITIC. **Chlorosis**. Manganese deficiency.

POWDERY MILDEW. **Uncinuliella australiana**, general; **Phyllactinia corylea**, AL.
ROT, Root. **Clitocybe tabescens**, FL; **Phymatotrichum omnivorum**, TX.
SOOTY MOLD. **Capnodium** spp., in aphid honeydew.

Mildew is serious on crape-myrtle, calling for sprays as the buds break.

CRASSULA
ANTHRACNOSE. **Gloeosporium** sp., NJ.
LEAF SPOT. **Phomopsis** sp., CT.
NEMATODE, Leaf. **Aphelenchoides** sp.
ROT, Root. **Armillaria mellea**, CA; **Pythium** sp., NJ, NY.

CREOSOTE BUSH (*Larrea*)
BLIGHT. **Omphalia** sp.
MISTLETOE. **Phoradendron californicum**, TX to CA.

CROWBERRY (*Empetrum*)
RUST. **Chrysomyxa empetri** (II, III), ME, NH, NY.

CRYPTOMERIA
BLIGHT, Leaf and Twig. **Diaporthe eres**; **Phyllosticta cryptomeriae**, VA and general U.S.
LEAF SPOT. **Pestalotia cryptomeriae**, SC; **P. funerea**, NJ.

CULVERS-ROOT (*Veronicastrum*)
LEAF SPOT. **Cercospora leptandrae**, WI; **Phyllosticta decidua**, TX, WI; **Ramularia veronicae**, TX; **Septoria veronicae**, WI.
POWDERY MILDEW. **Golovinomyces cichoracearum**, IL; **Sphaerotheca macularis**, CT, IL, IA, MD, MI, MO, WI.
ROT, Root. **Phymatotrichum omnivorum**, TX; **Rhizoctonia solani**, TX.
RUST. **Puccinia veronicarum** (III), IA, TX, WI.

CUNNINGHAMIA
LEAF SPOT. Necrotic. **Phomopsis** sp., OR.

CURRANT (*Ribes* spp.)
ANTHRACNOSE; Fruit Spot. **Pseudopeziza ribis**, general.
BLIGHT, Cane. **Bostryosphaeria ribis** var. **chromogena**, general.
BLIGHT, Gray Mold; Fruit Spot. **Botrytis cinerea**, Northeast, OR, WA.
BLIGHT, Thread. **Pellicularia koleroga**, FL.
CANKER, Cane Knot. **Thyronectria berolinensis**, CT to IN, KS, UT.
CANKER; DIEBACK. **Nectria cinnabarina**, ME to CO, WA; **N. ditissima**, MN, NY; Black Pustule, **Phragmodothella ribesia**, Northeast, Pacific Northwest.
DOWNY MILDEW. **Plasmopara viticola**, WI.
LEAF SPOT. **Alternaria** sp., MI; **Cylindrosporium ribis**, WI; **Mycosphaerella ribis**, ME to MD, AR, OR, WA.

LEAF SPOT, Angular. **Cercospora angulata**, NY to VA, KS, MN; **C. ribis**, AL, IN, IA.

POWDERY MILDEW. **Microsphaera grossulariae**, MT, NE, NH; **Sphaerotheca mors-uvae**, AK, CA, CT, MT, NE, OR, WA; **Phyllactinia corylea**, MI.

ROT, Berry. **Glomerella cingulata**, CT, PA.

ROT, Collar. **Fomes ribis**, NY to IN, MN, UT.

ROT, Root. **Armillaria mellea**, CA, OR, WA; **Phymatotrichum omnivorum**, TX; **Hypholoma perplexum**, NY.

RUST, White Pine Blister. **Cronartium ribicola** (II, III), ME to VA, IL, MN, OR, SD, WA. **Puccinia caricis** (0, I), CT, IN, MD, NY, II, III on Carex.

VIRUS. **Currant Mosaic**, MD, NY.

WILT. **Verticillum** sp., NY.

Black currant is the most important host of white pine blister rust and should not be grown; red currants should be kept at least 300 ft from pines. Cut out canes with cane blight.

CURRANT, FLOWERING (*Ribes*)

ANTHRACNOSE. **Pseudopeziza ribis**, IL, IA, MN, MT.

BACTERIAL Leaf Spot. **Pseudomonas ribicola**.

BLIGHT; DIEBACK. **Botrytis cinerea**, AK, IN.

CANKER; DIEBACK. **Botryosphaeria ribis**, KS; **Phragmodothella ribesia**, NY; **Thyronectria berolinensis**, KS; **Nectria cinnabarina**, AK, KS.

LEAF SPOT. **Cercospora angulata**, MN; **C. ribicola**, OR, WA; **Marssonina ribicola**, CO; **Mycosphaerella aurea**, NY to KS, SD, WA; **M. ribis**, NY to KS, MN, UT; **Phyllosticta grossulariae**, IN; **Septoria san-guinea**, WA.

RUST. **Coleosporium jonesii**, MN; **Cronartium occidentale**, MT to NM, CA, WA; **C. ribicola**, ME to MD, CO, MN, NV, Pacific states; **Melampsora ribesii-purpureae**; II, III on willow; **Puccinia caricina**, NY to IA, AK, CA, OR; **P. micrantha**; **P. parkerae**.

VIRUS. Viruslike vein-banding, CA, CT, MN, NJ, NY, OR, RI, WA.

CYPRESS (*Cupressus*)

BACTERIAL Crown Gall. **Agrobacterium tumefaciens**, AZ, CA, FL.

BLIGHT, Nursery. **Phomopsis juniperovora**, MD, NC, VA.

BLIGHT, Seedling. **Fusarium solani**, TX.

BLIGHT, Twig. **Coryneum asperulum**, AL; **C. berckmansii**, OR; **Cercospora thujina**, AL, GA, LA; **Pestalotia funerea**, CA, TX; **Botryosphaeria** sp., GA, on Arizona cypress.

CANKER. **Seiridium cardinale**, CA, Chile.

CANKER, Bark. **Seiridium unicorne**, CA.

CANKER; DIEBACK. **Coryneum cardinale**, on Monterey, sometimes Italian cypress; **Cytospora cenisia**, CA, on Italian cypress; **Macrophoma cupressi**, AL, CA, FL, TX; **Monochaetia unicornis**, GA, SC.

MISTLETOE. **Phoradendron** sp., AZ, CA, OR.

NEEDLE CAST. **Lophodermium** sp., NJ.
NEMATODE, Lesion. **Pratylenchus thornei**.
ROT, Heart. **Fomes pini**, CA.
ROT, Root. **Clitocybe tabescens**, FL; **Phymatotrichum omnivorum**, TX; **Phytophthora cinnamomi**, LA.
ROT, Wood. **Stereum taxodii**, cause of "pecky" cypress; **Coniophora puteana**, CA; **Lenzites saepiaria**, cosmopolitan; **Polyporus** spp.; **Poria subacida**; **Steccherinum ochraceum**.
RUST. **Gymnosporangium cupressi** (III), on Arizona cypress.

Coryneum canker has killed thousands of Monterey cypress trees in California, and Cytospora canker has been fatal in a narrow belt along the coast.

CYPRESS-VINE (*Quamoclit*)
NEMATODE, Root Knot. **Meloidogyne** sp., AL.
ROT, Root. **Phymatotrichum omnivorum**, TX.
RUST. **Coleosporium ipomoeae** (II, III), IL, SC, TN.
WHITE RUST. **Albugo ipomeae-panduratae**, MS, NM.

CYRILLA (Leatherwood)
CANKER, Brown Felt. **Septobasidium sinuosum**, FL.
LEAF SPOT. **Phyllosticta cyrillae**, FL.
RUST. **Aecidium cyrillae**, FL, LA, MS.

D

DAPHNE

BLIGHT, Twig. **Botrytis** sp., Northeast, Northwest.

DIEBACK; Wilt. **Fusarium** sp., NJ.

LEAF SPOT; Leaf Drop. **Gloeosporium mezerei**, WA; **Marssonina daphnes**.

ROT, Collar, Stem. **Phytophthora cactorum**, CA, NY; **Rhizoctonia solani**, NY.

ROT, Stem; Wilt. **Sclerotium rolfsii**, FL.

DATURA

BACTERIAL Canker, Vascular. **Clavibacter michiganense**, WY. On *Datura* sp., *D. innoxia*, *D. metal*, *D. meteloides*, *D. mollis*, *D. stramonium*.

BLIGHT, Southern. **Sclerotium rolfsii**, FL.

LEAF SPOT; Pod Blight. **Alternaria crassa**, FL, WI.

NEMATODE, Root Knot. **Meloidogyne hapla**.

ROT, Root. **Thielaviopsis basicola**, WI.

VIRUS. **Bromegrass Mosaic**, general; **Tomato Spotted Wilt**, CA, TX; **Potato Leafroll**, Pacific NW.

DECUMARIA

LEAF SPOT. **Cercospora decumariae**, MS.

DESERT BIRD OF PARADISE (*Caesalpinia*)

POWDERY MILDEW. **Leveillula taurica**, AZ.

DESERT-WILLOW (*Chilopsis*)

DAMPING-OFF. **Pythium ultimum**, NE; **Rhizoctonia solani**, NE, TX.

LEAF SPOT. **Phyllosticta erysiphoides**.

ROT, Root. **Phymatotrichum omnivorum**, TX.

DEUTZIA

LEAF SPOT. **Cercospora deutziae**, DE, IA, TX; **Phyllosticta deutziae**, AL, IA, NJ, TX.

NEMATODE, Root Knot. **Meloidogyne** spp.

ROT, Root. **Armillaria mellea**, CA.

R.K. Horst, *Field Manual of Diseases on Trees and Shrubs*,
DOI 10.1007/978-94-007-5980-0_7, © Springer Science+Business Media Dordrecht 2013

DEVILS-CLUB (*Oplopanax*)
BLIGHT, Gray Mold. **Botrytis cinerea**, AK.
LEAF SPOT. **Cercospora daemonicola**, OR.

DEVILWOOD (*Osmanthus americanus*)
BLACK MILDEW. **Asterina asterophora**, FL, GA; **A. discoidea**; **A. purpurea**, FL; **Lembosiaoleae**, MS; **Meliola amphitricha**, FL to MS.
LEAF SPOT. **Phyllosticta oleae**, FL, NC.
MISTLETOE. **Phoradendron serotinum** (**flavescens**), FL.

DEWBERRY (*Rubus*)
BACTERIAL Cane Gall. **Agrobacterium rubi**, NY, OR.
BACTERIAL Crown Gall. **Agrobacterium tumefaciens**, general.
BACTERIAL Hairy Root. **Agrobacterium rhizogenes**, OR.
BLACK MILDEW. **Irenina sanguinea**, AL, LA.
BLIGHT, Cane; DIEBACK. **Leptosphaeria coniothyrium**, general; **L. thomasiana**, OR, WA.
BLIGHT, Spur. **Didymella applanata**, OR, WA.
BLIGHT, Stamen; Dry Berry. **Hapalosphaeria deformans**, OR, WA.
BLOTCH, Sooty. **Gloeodes pomigena**, NC, PA.
CANKER; Fruit Rot. **Glomerella cingulata**, GA, IL, MD, MS.
CANKER; Cane Spot. **Ascospora ruborum**, AL, CA.
DOWNY MILDEW. **Peronospora rubi**, FL, MD, WI, WA; **P. potentillae**, CT, IL, LA.
FRUIT SPOT; Flyspeck. **Leptothyrium pomi**, NC.
GALL, Yellow Leaf. **Synchytrium aureum**, WI.
LEAF SPOT. **Mycosphaerella confusa**, NJ to FL, IL, TX; **Pezizella oenotherae**, MD to NC; **Phyllosticta ruborum**, NY; **P. dispergens**, IL; **Septoria rubi** (**Mycosphaerella rubi**), general; **S. darrowi**, NY.
POWDERY MILDEW. **Sphaerotheca macularis**, IL, IN, MN, OH, PA, Pacific Northwest.
ROSETTE; Double Blossom. **Cercosporella rubi**, NC to AL, LA, MS, TX, IL.
ROT, Collar. **Rhizoctonia solani**, TX, WA.
ROT, Fruit. **Botrytis cinerea**; **Phyllosticta carpogena**, MD, NJ, NY, NC.
ROT, Root. **Armillaria mellea**, OR; **Collybia dryophila**, NC; **Corticium galactinum**, MD, VA, TX; **Helicobasidium purpureum**, NC, TX; **Phymatotrichum omnivorum**, TX.
RUST. **Gymnoconia peckiana**, orange (0, I, III), ME to VA, CA, MN, MO; **Kuehneola uredinis**, yellow (0, I, II, III), ME to FL, CA, KS, TX, WA; **Kunkelia nitens**, orange (I), CA, OR, CT to FL, IA, TX; **Mainsia rubi** (II, III), TX.
SPOT ANTHRACNOSE. **Elsinoë veneta**, general.
VIRUS. **Raspberry Beta Leaf Curl**, MI, OH, TX; **Raspberry Mosaic**, CT, MI, NJ, NY, Pacific Coast; **Loganberry Dwarf**, especially on loganberry, CA, OR, WA.
WILT. **Verticillium albo-atrum**, CA, OR, WA.

Use virus-free, bacterial-free, planting stock; spray for anthracnose.

DIEFFENBACHIA

Anthracnose. **Glomerella cincta**, NJ; **Gloeosporium** sp., WA; **Colletotrichum** sp., WA.

Bacterial Stem and Leaf Rot. **Erwinia dieffenbachiae**, FL.

Leaf Spot. **Cephalosporium dieffenbachiae**, FL, NY; **Leptosphaeria** sp., FL; **Myrotheciumroridum**, FL.

Rot, Root. **Pythium splendens**, FL; **Rhizoctonia** sp., FL.

Rot, Stem. **Phytophthora palmivora**, CA, FL.

Rot, Stem, Leaf, Cutting. **Fusarium solani**, FL.

Propagate from disease-free canes.

DIERVILLA (Bush Honeysuckle)

Leaf Spot. **Cercospora weigeliae** (*C. diervillae*) ME, TX; **Phyllosticta diervillae**, WI; **Ramularia diervillae**, ME, NH, NY, WI; **Septoria diervillae**, IA, MA, MN, WI.

Nematode, Root Knot. **Meloidogyne** sp.

Powdery Mildew. **Microsphaera alni**, NJ, WA.

Rot, Root. **Phymatotrichum omnivorum**, TX.

Virus. **Tobacco Ringspot**, MD.

DIODIA

Virus. **Diodia Vein Chlorosis**, AR.

DOGWOOD, DWARF, BUNCHBERRY (*Cornus canadensis*)

Leaf Spot. **Ceratobasidium anceps**, rot, NH; **Discohainesia oenotherae**, also stem spot, ME, NH; **Phyllosticta** sp., NY; **Ramularia** sp., NY; **Septoria canadensis**, AK, ME, WA; **Glomerularia corni**, ME to WI, OR.

Powdery Mildew. **Phyllactinia corylea**, WA.

Rust. **Puccinia porphyrogenita** (III), ME to AK, WA.

DOGWOOD, FLOWERING (*Cornus florida*)

Anthracnose. **Colletotrichum gloeosporioides**, TX; **Discula** sp., ID, OR, Pacific NW, MD; **D. destructive**, CT, GA, NC, TN.

Bacterial Crown Gall. **Agrobacterium tumefaciens**, MO.

Bacterial Leaf Blight. **Pseudomonas syringae**, TN.

Black Mildew. **Dimerosporium pulchrum** and **Meliola nidulans**, Southeast.

Blight, Flower and Leaf. **Botrytis cinerea**, MD, MA, NJ, NY, probably general in wet springs.

Blight, Foliar. **Phytophthora parasitica**, FL; P. sp., TN.

Blight, Thread. **Pellicularia koleroga**, LA.

Blight, Twig, Fruit Spot. **Colletotrichum acutatum**, CT, FL.

Canker, Crown; Bleeding; Collar Rot. **Phytophthora cactorum**, MD, MA, NJ, NY, WA.

Canker; Dieback. **Botryosphaeria ribis**, PA; **Cytospora** sp., NJ; **Cryptostictis**, NY; **Sphaeropsis** sp.

CANKER, Felt Fungus. **Septobasidium** spp., VA to FL, LA; **Cryptodiaporthe corni** Anamorph, **Myxosporium nitidum**, ND.
DAMPING-OFF. **Colletotrichum acutatum** and **Fusarium oxysporum**, GA.
LEAF SPOT. **Cristulariella pyramidalis**, FL; **Phyllosticta cornicola**, PA to VA, KS, TN; **Ascochyta cornicola**, NC, PA; **Cercospora cornicola**, NC to FL; **Septoria cornicola**, NY to GA, IA; **Colletotrichum gloeosporioides**, NY.
MISTLETOE. **Phoradendron serotinum (flavescens)**, FL.
NEMATODE, Dagger. **Xiphinema americanum**.
NEMATODE, Lance. **Hoplolaimus uniformis**, RI.
NEMATODE, Sting. **Belonolaimus longicaudatus**, FL.
NONPARASITIC. **Scorch**. Water deficiency, frequent in Southeast.
POWDERY MILDEW. **Microsphaera alni**, MA to NC, IL, WI; **M. pulchra**, CT, SC; **Phyllactinia corylea**, general.
ROT, Root. **Armillaria mellea**, NY; **Clitocybe tabescens**, FL, GA; **Phymatotrichum omnivorum**, TX; **Corticium galactinum**, VA; **Pythium** sp.
ROT, Wood. **Daedalea confragosa**, MD, NC, PA, TN, VA, WV; **Daldinia vernicosa**, MD; **Lenzites betulina**, NC; **Polyporus** spp.; **Poria** spp., MD, PA.
SPOT ANTHRACNOSE. **Elsinoë corni**, DE, FL, GA, LA, MD, NC, SC, VA.
VIRUS. **Tobacco Ring Spot**, MD. **Witches' Broom Disease**, NJ; **Dogwood Mosaic**, SC.
WILT. **Verticillium** sp., MA.

The most serious dogwood disease in the East is crown or bleeding canker, which attacks trees after transplanting or injury. In a wet season spot anthracnose badly disfigures leaves, twigs, berries. In wet weather Botrytis blight is conspicuous as flowers fade and petals rot onto leaves.

DOGWOOD, PACIFIC (*Cornus nuttalli*)
CANKER, Bleeding; Collar Rot. **Phytophthora cactorum**, WA.
CANKER, Trunk. **Nectria galligena**, OR, WA.
POWDERY MILDEW. **Phyllactinia corylea**, general.
ROT, Root. **Armillaria mellea**, WA; Heart, **Fomes igniarius**, OR.

DOGWOOD, PAGODA, GRAY, RED OSIER, WESTERN OSIER (*Cornus* spp.)
Most of the diseases listed for flowering dogwood occur on these shrub dogwoods.

DOUGLAS-FIR (*Pseudotsuga*)
BACTERIAL Gall. **Agrobacterium pseudotsugae**, CA.
BLIGHT, Brown Felt. **Herpotrichia nigra**, Rocky Mts. and Pacific Northwest.
BLIGHT, Gray Mold; Snow Mold. **Botrytis cinerea**, cosmopolitan.
BLIGHT, Needle. **Rosellinia herpotrichioides**, CA.
BLIGHT. **Phoma eupyrena**, CA; **Phytophthora ramorum** (also small branch canker), CA.
BLIGHT, Seedling. **Rhizina undulata**, Pacific Northwest; **Fusarium oxysporum**, OR.
BLIGHT, Seedling Smother. **Thelephora terrestris**, Pacific Northwest.
BLIGHT, Snow. **Phacidium infestans**, ID.

CANKER, Bark. **Brunchorstia** (*Cryptosporium*) **boycei**, WA; **Chondropodium pseudotsugae**, OR; **Aleurodiscus** spp., weakly parasitic.

CANKER, Branch. **Dermea pseudotsugae**, CA.

CANKER, Branch, Trunk. **Dasyscyphus pseudotsugae**; **D. ellisiana**, twig, MA, NC, RI; **Phomopsis lokoyae**, Pacific Coast; **Cytospora** sp., twig, CO, NJ, OR; **Phaciopycnis** (*Phomopsis*) **pseudotsugae**; **Phomopsis lokoyae**, CA.

CANKER; DIEBACK. Collar Rot; Seedling Blight. **Diplodia pinea**, CA, KS, NJ, NY.

DAMPING-OFF. **Rhizoctonia solani**, cosmopolitan.

MISTLETOE; Witches Broom. **Arceuthobium douglasii**, MT to CO, OR, WA. **Phoradendron serotinum (flavescens)**, TX.

NEEDLE CAST. **Adelopus gaeumannii**, Pacific states, Northeast; **Rhabdocline pseudotsugae**, general; **R. pseudotsugae** subsp. **pseudotsugae**, PA; **R. weirii**, PA; **Rhabdogloeum hydrophyllum**, AZ, NM.

NEEDLE CAST, Swiss. **Phaeocryptopus gaeumannii**, general, MN (also Lake states).

NEEDLE, Flyspeck. **Leptothyrium pseudotsugae**, CO.

NEMATODE, Ring. **Criconemoides crotaloides**; **Meloidodera** sp., OR.

ROT, Root. **Armillaria mellea**, cosmospolitan; **Sparassis radicata**, Pacific Northwest; **Phytophthora cinnamomi**, OR, WA, Southeast; **Phymatotrichum omnivorum**, TX; **Verticicladiella wagenerii**, CA, MT.

ROT, Root. **Armillaria mellea**, PA; **Fusarium solani**, PA; **F. oxysporum**, PA; **F. avenaceum**, PA.

ROT, Stem, Seedling. **Fusarium avenaceum**, OR; **sambucinum**, OR.

ROT, White Pocket. **Ganoderma oregonensis**, MT, OR, WA; **Hydnum coralloides**, Pacific Northwest.

ROT, Heart, Wood. **Poria weirii**, destructive, OR, WA; **Echinodontium tinctorium**; **Fomes** spp.; **Polyporus** spp.; **Lenzites saepiaria**, wide-spread; **Stereum** spp.; **Trametes** spp.

RUST. **Melampsora albertensis** (0, I), MT to CO, UT, WA; II, III on poplar.

RUST, Leaf. **Melampsora medusae** f. sp. **deltoidae** (O, I), WA; II, III on poplar.

DURANTA

BLIGHT, Seedling. **Sclerotium rolfsii**, FL.

LEAF SPOT, Black. **Phyllachora fusicarpa**, FL.

E

ELAEAGNUS (Russian-Olive, Silverberry)

BACTERIAL Crown Gall. **Agrobacterium tumefaciens**, GA; Hairy Root, **A. rhizogenes**, IA.

BLIGHT, Southern. **Sclerotium rolfsii**, TX.

BLIGHT, Stem and Branch, Bark and Cambium Necrosis. **Botryodiplodia theobromae**, ND, SD, Southern Plains, NE, Great Plain states.

BLIGHT, Thread. **Rhizoctonia ramicola**, FL.

BLIGHT, Tip. **Gloeosporium fructigenum**.

CANKER. **Nectria cinnabarina**, CA; **Fusarium** sp., WY; **Phytophthora cactorum**, AZ, IL; **Fusicoccum elaeagni**, IL; **Tubercularia ulmea**, ND, SD.

CANKER, Phomopsis. **Phomopsis elaeagni**, (*Fusicoccum elaeagni*) DE, OH; **P. arnoldia** (syn. *P. elaeagni*), MI, ND, SD.

LEAF SPOT. **Cercospora carii**, TX; **C. elaeagni**, FL, MS, OK, TX; **Phyllosticta argyrea**, MD, NC; **Septoria argyrea**, IA, NE, ND, WI; **S. elaeagni**, KS.

MISTLETOE. **Phoradendron serotinum (flavescens)**, FL.

POWDERY MILDEW. **Phyllactinia corylea**, OR; **P. elaeagni**, WY.

ROT, Root. **Clitocybe tabescens**, FL.

RUST. **Puccinia caricis-shepherdiae** (0, I), Northern Plains; II, III on *Carex*; **P. coronata** (0, I), MT, ND; II, III on *Calamagrostis*.

WILT. **Verticillium** sp., WA.

ELDER (*Sambucus*)

BLIGHT, Thread. **Pellicularia koleroga**, LA; Web, **P. filamentosa**, FL.

CANKER, Branch. **Cytospora sambucicola**, IL; **C. chrysosperma**.

CANKER; DIEBACK. **Botryosphaeria ribis**, FL, GA; **Diplodia** spp.; **Nectria cinnabarina**, widespread; **N. coccinea**, MD, MI, WA; **Sphaeropsis sambucina**.

GALL, Leaf. **Synchytrium sambuci**, LA.

LEAF SPOT. **Ascochyta sambucina**, AK; **A. wisconsina**, NY, WI; **Cercospora catenospora**, AL, KS, MS; **C. depazeoides**, general; **Cercosporella prolificans**, CA, NM, OR; **Gloeosporium tineum**, MS, TX; **Phyllosticta sambuci**, MO,

R.K. Horst, *Field Manual of Diseases on Trees and Shrubs*,
DOI 10.1007/978-94-007-5980-0_8, © Springer Science+Business Media Dordrecht 2013

NY, WI; **Ramularia sambucina**, MO, NY, WI, WA; **Mycosphaerella** sp., NM; **Septoria sambucina**, VT to FL, CA, OR, TX, WA.

POWDERY MILDEW. **Microsphaera alni** and **M. grossulariae**, general; **Phyllactinia corylea**, MI; **Sphaerotheca macularis**, MA.

ROT, Heart, Wood. **Fomes igniarius**, ID; **Hymenochaete agglutinans**, WY; **Polyporus** spp.

ROT, Root. **Helicobasidium purpureum**, TX; **Phymatotrichum omnivorum**, TX; **Xylaria multiplex**, TX.

RUST. **Puccinia bolleyana** (0, I), ME to FL, MN, TX; II, III on *Carex*.

SPOT ANTHRACNOSE. **Sphaceloma** sp., LA.

VIRUS. **Tobacco Mosaic**.

WILT. **Verticillium albo-atrum**, MD.

ELDERBERRY

VIRUS. **Tobacco Ring Spot**.

ELM (*Ulmus*)

BACTERIAL, Leaf Scorch. **Xylella fastidiosa**, OK.

BACTERIAL, Leaf Spot. **Pseudomonas syringae**, PA.

BACTERIAL, MLO. **Phloem Necrosis**, AL, AR, GA, IL, IN, IA, KS, KY, MS, MO, NE, NJ, NY, OH, OK, PA, TN, WV.

BACTERIAL Wetwood. **Erwinia nimipressuralis**, VA, on Siberian elm (*Ulmus pumila*).

BACTERIAL Wetwood; Slime Flux. **Erwinia nimipressuralis**.

BLIGHT, Twig. **Septogloeum parasiticum**, MI; **Phomopsis oblonga**, MA; **Fusarium** spp.

BLIGHT, Twig. **Phomopsis oblonga**, VA, on Chinese elm (*Ulmus parvifolia*).

CANKER, Bleeding. **Phytophthora cactorum**, RI; Pit, **P. inflata**, CT, MA, NY, PA.

CANKER, Felt Fungus. **Septobasidium pseudopedicellatum**, NC.

CANKER, Stem. **Botryosphaeria ribis**.

CANKER, Twig. **Sphaeropsis ulmicola**, VA, on Chinese elm (*Ulmus parvifolia*).

CANKER, Twig Dieback. **Apioporthe apiospora**, IA; **Coniothyrium** spp., IL, MA, MI; **Cytospora ludibunda**, CT, PA; **Nectria coccinea**, NJ, NY; **N. cinnabarina**, coral spot, widespread; **Phoma** sp., **Phomopsis** sp., Northeast to IL, MN, SC; **Sphaeropsis** sp., CT to MS; **Cytosporina ludibunda**, IL, KS.

DAMPING-OFF. **Rhizoctonia solani**, cosmopolitan; **Pythium** sp.

DIEBACK. **Cephalosporium** sp., VA, on English Elm (*Ulmus procera*).

LEAF BLISTER. **Taphrina ulmi**, CT to MS, MO, WI.

LEAF SPOT. **Cercospora sphaeriaeformis**, LA, TX; **Cylindrosporium tenuisporium**, TX; **Coryneum tumoricola**, NY; **Gloeosporium ulmicola**; **Monochaetia desmazierii**, GA; **Phyllosticta confertissima**, PA; **Mycosphaerella ulmi**, MA to AL; **Septogloeum profusum**, AL; **Coniothyrium ulmea**, WV; **Ceratophorum ulmicola**, KS, NE.

LEAF SPOT; Anthracnose. **Gloeosporium inconspicuum**, MA to VA; MN, OK.

LEAF SPOT; Black Spot. **Gnomonia ulmea**, general.

MISTLETOE. **Phoradendron serotinum** (**flavescens**), IN, TX; **P. tomento-sum**, TX.

Mistletoe, European. **Viscum album**, CA.
Nematode, Dagger. **Xiphinema americanum.**
Nematode, Root Knot. **Meloidogyne** spp.
Nematode, Stem. **Ditylenchus gallicus**; Leaf, **Aphelenchoides fragariae**.
Phytoplasma. **Elm Yellows**, NY.
Powdery Mildew. **Microsphaera alni**, IL, IA, MS, OH; **Phyllactinia corylea**, NC to TX, IA; **P. ungulata**, GA; **Uncinula macrospora**, general.
Rot, Heart. **Collybia velutipes**, widespread; **Daedalea confragosa**, wide-spread; **Fomes** spp.; **Ganoderma curtisii**, NY.
Rot, Root. **Helicobasidium purpureum**, TX; **Phymatotrichum omnivorum**, TX; **Xylaria** spp.; **Armillaria mellea**, MO; **Clitocybe tabescens**, FL.
Rot, Wood. **Daldinia concentrica**, widespread; **Lenzites betulinum**, IN, MA, MD; **Pleurotus ostreatus**, widespread; **Polyporus** spp.; **Schizophyllum commune**, cosmopolitan; **Ustulina vulgaris**, MD.
Wilt. **Dothiorella** (*Cephalosporium*) **ulmi**, general; **Verticillium alboatrum**, ME to VA, MS, OR, WI.
Wilt; Dutch Elm Disease. **Ceratocystis ulmi**, general.

The Dutch elm disease and phloem necrosis have taken a heavy toll of elms in many states. A dormant spray for the bark beetles that spread Dutch elm disease is the present recommendation, combined with general sanitation. Chemotherapy is still promising but not yet practical. Some seedling elms are highly resistant but not immune.

ENCELIA
Nematode, Root Knot. **Meloidogyne** sp., CA.
Rust. **Puccinia enceliae** (0, I, III), CA.

EPIGAEA (Mayflower, Trailing Arbutus)
Leaf Spot. **Cercospora epigaeae**, NY, NC, WI; **Phyllosticta epigaeae**, MA, NY.
Powdery Mildew. **Microsphaera alni** var. **vaccinii**, CT to VA, WI.

EPILOBIUM (Willow-Herb, Fireweed)
Bacterial, Mlo. **California Aster Yellows**, CA.
Blight, Gray Mold. **Botrytis cinerea**, AK.
Blight, Southern. **Sclerotium rolfsii**, TX.
Downy Mildew. **Plasmopara epilobii**, AK, IL, NY.
Leaf Spot. **Cercospora montana**, widespread; **Discosia bubaki**, NY, WI; **Phyllosticta chamaeneri**, OR; **P. wyomingensis**, WY; **Ramularia cercosporoides**, AK, MT, TX, WA, WY; **Septoria epilobii**, CA, DE, IL, VT, WI.
Nematode, Root Knot. **Meloidogyne** sp.
Powdery Mildew. **Sphaerotheca macularis**, widespread; **Erysiphe polygoni**, WA; **Microsphaera** sp., IL.
Rot, Root. **Phymatotrichum omnivorum**, TX.
Rust. **Puccinia epilobii** (III), MI, WY; **P. dioicae** (0, I), CO; **P. gigantea** (III), ID, MT, TX, WA, WY; **P. scandica** (III), UT, WA, WY; **P. oenotherae** (0, I, II, III),

CA; **P. pulverulenta** (0, I, II, III), ND to NM, CA; **P. veratri** (0, I), NH, MT to WA; **Pucciniastrum pustulatum** (II, III) widespread; 0, I on fir.
SMUT, Leaf. **Doassansia epilobii**, CO, NH.

ERANTHEMUM
LEAF SPOT. **Phyllosticta** sp., NJ.

ERYTHRINA
BLIGHT, Thread. **Pellicularia koleroga**, FL; **Rhizoctonia ramicola**, FL.
NEMATODE, Root Knot. **Meloidogyne** sp.
ROT, Root. **Clitocybe tabescens**, FL; **Phymatotrichum omnivorum**, TX.
RUST, Soybean. **Phakopsora pachyrhizi**, FL.
WILT. **Verticillium albo-atrum**, CA.

ERYTHROXYLUM
WILT. **Fusarium oxysporum** f. sp. **erythroxyli**, HI.

EUCALYPTUS (Gum-Tree)
BACTERIAL Crown Gall. **Agrobacterium tumefaciens**, CA.
BLIGHT, Seedling. **Fusarium oxysporum** f. sp. **aurantiacum**, CA.
CANKER. **Diaporthe cubensis**, FL, HI, PR; **Lasiodiplodia pseudotheobromae, Neofusicoccum eucalyptorum, N. parvum, N. ribis**, Uruguay.
CANKER; DIEBACK, Seed Capsule Abortion. **Botryosphaeria ribis**, FL.
CANKER, Felt Fungus. **Septobasidium curtisii**, NC.
LEAF SPOT. **Actinopelte dryina**, LA; **Mycosphaerella molleriana**, CA; **Phyllosticta extensa**, CA.
NONPARASITIC. **Chlorosis**. Iron deficiency, CA.
POWDERY MILDEW. **Golovinomyces cichoracearum**, AZ.
ROT, Heart; Wood. **Ganoderma applanatum**; **Polyporus** spp.; **Stereum hirsutum**.
ROT, Root. **Clitocybe tabescens**, FL; **Armillaria mellea**, CA; **Phymatotrichum omnivorum**, TX.

Many other fungi may be found on leaves, twigs, and branches but are not reported as causing specific diseases.

EUGENIA
BLACK MILDEW. **Asterinella puiggarii**, FL.
DIEBACK. **Fusicoccum, Neoscytalidium** and **Botryosphaeria**, FL.
LEAF SPOT. **Pezizella oenotherae**, NY.
ROT, Root. **Clitocybe tabescens**, FL.

EUONYMUS (Burning-Bush, Spindle-Tree)
ANTHRACNOSE. **Colletotrichum griseum**, AL, NJ, NY, GA to TX, AR; **Gloeosporium frigidum**, AR, MS.
BACTERIAL Crown Gall. **Agrobacterium tumefaciens**, CT, MI, MS, NJ, SC, TX.
BLIGHT, Thread. **Pellicularia koleroga**, LA.
DIEBACK, CANKER, Basal. **Whetzelinia sclerotiorum** (*Sclerotinia*), RI.

LEAF SPOT. **Cercospora destructiva**, VA to TX; **C. euonymi**, PA to WI; **Exosporium concentricum**, AL, MS, SC, TX, VA; **Phyllosticta euonymi**, NY to MS and TX; **P. pallens**, AL; **Septoria euonymi**, MS, VA; **S. atropurpurea**, IL; **Ramularia euonymi**, CA, IA, KS, MO; **Marssonina thomasiana**, OH to WI, MO; **Myocentrospora** sp., OH.

NEMATODE, Root Knot. **Meloidogyne** sp., MD, TX.

POWDERY MILDEW. **Microsphaera alni**, NJ to SD and southward; **Oidium euonymi-japonici**, AL, CA, IA, LA, MS, NJ, SC, TX, WA.

ROT, Crown. **Phytophthora cactorum**.

ROT, Root. **Phymatotrichum omnivorum**, TX; **Fusarium scirpi**, NJ.

VIRUS. **Euonymus Mosaic**; Infectious Variegation, MA.

The Oidium mildew is prevalent throughout the South and in California. Crown gall is common, with conspicuous knobs along the vines, but seldom fatal.

EURYOPS

POWDERY MILDEW. **Podosphaera** (Shaerotheca) **jusca**, CA.

F

FARKLEBERRY, TREE-HUCKLEBERRY (*Vaccinium*)

CANKER, Felt Fungus. **Septobasidium sinuosum**, FL.

GALL, Leaf. **Exobasidium vaccinii**, AL, FL.

LEAF SPOT. **Cylindrosporium** sp., TX; **Ophiodothella vaccinii**, MD to GA, TX; **Pestalotia vaccinicola, secondary**, FL; **Phyllosticta vaccinii**, AL, FL, MS, TX; **Septoria albopunctata**, FL, SC, TX; Tar Spot, **Rhytisma vaccinii**, FL, OK, TX.

ROT, Root. **Phymatotrichum omnivorum**, TX.

VIRUS. **Blueberry Scorch**, British Columbia, Canada.

FEIJOA (*Pineapple Guava*)

BLIGHT, Thread. **Rhizoctonia ramicola**, FL.

ROT, Fruit. **Botrytis cinerea**, CA; **Colletotrichum gloeosporioides**, CA; **Penicillium expansum**, CA.

ROT, Root. **Phymatotrichum omnivorum**, TX.

SPOT ANTHRACNOSE. **Sphaceloma psidii**, FL.

FENDLERA

RUST. **Gymnosporangium speciosum** (0, I), CO and UT to NM, AZ; II, III on juniper.

FIG (*Ficus carica*)

ANTHRACNOSE; Fruit Rot. **Colletotrichum gloeosporioides** (*Glomerella cingulata*), NC to TX.

BACTERIAL Crown Gall. **Agrobacterium tumefaciens**, CA, FL, TX.

BLOTCH, Leaf. **Cercospora fici**, NC to FL, TX.

BLIGHT, Limb. **Corticium salmonicolor**, Gulf states.

BLIGHT, Southern. **Sclerotium rolfsii**, FL.

BLIGHT, Thread. **Pellicularia koleroga**, FL, LA, MS; Web, **P. microsclerotia**, FL to LA, TX.

BLIGHT, Twig. **Gibberella baccata**, CA.

R.K. Horst, *Field Manual of Diseases on Trees and Shrubs*,

DOI 10.1007/978-94-007-5980-0_9,

CANKER; DIEBACK. **Botryosphaeria ribis**; **Diplodia sycina**, NC, OR; **Macrophoma fici**, also fruit dry rot, LA, NC, TX; **Nectria cinnabarina**, TX; **Megalonectria pseudotrichia**, LA, TX; **Physalospora rhodina**, AL, FL, TX; **Sclerotinia sclerotiorum**, CA, TX.

CANKER, Felt Fungus. **Septobasidium** spp., MS.

CANKER, Phomopsis. **Phomopsis cinerescens**, CA.

LEAF SPOT. **Colletotrichum elastica**, FL; **Corynespora cassiicola**, FL; **Ascochyta caricae**, CA; **Alternaria** sp., SC; **Cephalosporium acremonium**, LA; **Eutypa** sp., TX; **Mycosphaerella bolleana**, NC to TX; **Ormathodium fici**, LA.

NEMATODE, Dagger. **Xiphinema index**, CA.

NEMATODE, Fig Cyst. **Heterodera fici**, CA.

NEMATODE, Lesion. **Pratylenchus vulnus**, **P. musicola** and **P. pratensis**, CA.

NEMATODE, Pin. **Paratylenchus hamatus**, CA.

NEMATODE, Root Knot. **Meloidogyne** spp., southern states to CA.

NONPARASITIC. **Chlorosis**. Manganese deficiency, FL.

Little Leaf. Zinc deficiency, CA.

Sunburn. Low temperature injury to trunk and branches, CA.

ROT, Fruit. **Alternaria tenuis**, in market; **Aspergillus niger**, black mold; **Botrytis cinerea**; **Choanephora cucurbitarum**, TX; **Cladosporium herbarum**, market; **Diplodia natalensis**, TX; **Fusarium moniliforme**, CA; **Oospora** sp., sour rot, TX; **Rhizopus nigricans**, CA, Gulf states; **Trichothecium roseum**, Gulf states.

ROT, Root. **Armillaria mellea**, CA; **Dematophora necatrix**, NC; **Phymatotrichum omnivorum**, TX; **Rhizoctonia** sp., TX.

RUST. **Cerotelium** (*Physopella*) **fici** (II, III), NC to FL, AR, TX.

SOOTY MOLD. **Capnodium** sp., TX; **Fumago vagans**, Gulf states.

VIRUS. **Fig Mosaic**, AZ, CA, GA, TX, VA.

WILT, Branch. **Hendersonula toruloides**, CA.

FIG, FLORIDA STRANGLER (*Ficus aurea*)

ANTHRACNOSE. **Colletotrichum gloeosporioides**, FL.

BACTERIAL Crown Gall. **Agrobacterium tumefaciens**, FL.

LEAF SPOT. **Ophiodothella fici**, FL; **Phlyctema ficuum**, FL; **Phyllosticta physopellae**; **P. roberti**, FL.

NEMATODE, Root Knot. **Meloidogyne** spp.

ROT, Fruit. **Fusarium moniliforme**, CA.

RUST. **Cerotelium fici** (II, III), FL.

FILBERT

See Hazelnut.

FIR (*Abies*)

BLACK MILDEW. **Dimerosporium abietis**, ID, OR, WA; **Adelopus nudus**, ME, NY, NC, WI.

BLIGHT. **Phoma eupyrena**, CA.

BLIGHT, Brown Felt. **Herpotrichia nigra**, on high western firs.

Blight, Needle. **Acanthostigma parasiticum**; **Rehmiellopsis balsameae**, Northeast; **Cenangium ferruginosum**, MI, PA; **Rhabdogloeum abietinum**, NC; **Macrophoma parca**, CO, ID, MT, OR, WA; **Phoma eupyrena**, CA; **Furcaspora pinicola**, CA; **Phyllosticta multicorniculata**, NH.

Blight, Needle and Branch. **Phytophthora capsici**, MI.

Blight, Seedling Smother. **Thelephora terrestris**.

Blight, Snow. **Phacidium infestans**, Northeast; **P. balsameae**; **P. abietinellum**.

Blight, Tip. **Delphinella balsameae**, VT.

Canker. **Aleurodiscus amorphus**, general; **Cytospora pinastri**, ME, WI **C. cylindroides** and **C. abietis**, OR, WA; **Cephalosporium** sp., MN, WI; **Cryptosporium macrospermum**, New England; **Dasyscyphus resinaria**, MN; **Nectria fuckeliana**, CA, OR; **N. balsameae**, NC, PA; **Phomopsis boycei**, ID, MT; **P. fokoyae**, CA.; **P. montanensis**; **Ophionectria scolecospora**, widespread; **Scleroderris abieticola**, OR; **Sphaeropsis abietis**, MI; **Valsa** (*Cytospora*) **kunzei**; **Thyronectria balsamea**, CO.

Canker, Pitch. **Fusarium circinatum**, CA.

Dieback, Twig. **Sydowia polyspora**, CA.

Mistletoe. **Arceuthobium americanum**, WA; **A. campylopodium**, widespread in West; **A. douglasii**, NM, OR, WA; **A. laricis**, WA; **Phoradendron pauciflorum**, AZ, CA.

Needle Cast; Blight. **Bifusella abietis**, ID, MT, WA; **B. faullii**, ME, MI, NH; **Hypoderma robustum**, CA, ID, OR, WA; **Hypodermella abietis-concoloris**, widespread on western firs; **H. mirabilis**, MI; **H. nervata**, ME, NH, VT; **H. punctata**, ID, OR; **Lophodermium piceae**, tar spot, widespread, weakly parasitic; **L. autumnale**, CA, ID, MI, OR, UT, WY; **L. consciatum**; **L. decorum**; **L. uncinatum**; **L. lacerum**, NH, NY, PA, VT.

Needle Cast; Blight. **Tirula nervisequa conspicuous**, CA.

Nematode. **Nacobbodera chitwoodi**, OR.

Rot, Heart. **Stereum sanguinolentum**, widespread; **Polyporus sulphureus**, general.

Rot, Crown. **Phytophthora cactorum**, VA.

Rot, Hypocotyl. **Fusarium oxysporum**, CA.

Rot, Root. **Armillariamellea**, New England, NY; **Fusarium solani**, PA; **F. oxysporum**, PA; **F. avenaceum**, PA; **Phytophthora cinnamomi**, NC; **P. capsici**, MI; **P. citricola**, NC; **P. drechsleri**, NC; **P. ramorum**, CA; **Verticicladiella wagenerii**, Pacific Coast; **Poria weirii**, OR.

Rot, Wood. **Coniophora puteana**, New England, NY; **C. corrugis**, AZ, CO, WY; **Fomes annosus**, CO; **Echinodontium tinctorium**; **Hydnumabietis**; **H. balsameum**; **Polyporus** spp.; **Poria** spp.

Rust, Fir-Broom. **Melampsorella caryophyllacearum, NY**.

Rust, Fir-Fern. **Milesia fructuosa** (0, I) on new needles, ME, NH, NY; II, III on *Dryopteris*; **M. marginalis**, MA, NH, NY; **M. polypodophila** (0, I), ME, NH, NY, on old needles; II, III on *Polypodium*, **Hyalopsora aspidiotus** (0, I), on 2-year needles; II, III on *Dryopteris*; **Uredinopsis mirabilis** (0, I), general; II,

III on sensitive fern; **U. osmundae** (0, I), widespread; II, III on *Osmunda*; **U. struthiopteridis**, ID, MI, OR, WA; II, III on ostrich-fern; **U. phegopteridis** (0, I); II, III on *Dryopteris*.

RUST, Fir-Fireweed. **Pucciniastrum pustulatum**, widespread; II, III on Epilobium.

RUST, Fir-Huckleberry. **Pucciniastrum goeppertianum**, widespread on western firs, and ME, PA, WI; II, III on *Vaccinium*.

RUST, Fir-Willow. **Melampsora abieti-capraearum**, widespread; II, III on willow.

RUST, Needle. **Caeoma faulliana**, OR; **Peridermium ornamentale**, ID, MT, OR, WA; **P. rugosum**, CA, OR, WA.

RUST, Witches' Broom. **Melampsorella cerastii**, general; II, III on chickweed.

WILT, Dieback. **Phytophthora ramorum**, CA.

Despite this long list of possibilities the gardener should not have much trouble with ornamental firs. Rehmiellopsis tip blight yellows needles of new growth and causes twig dieback of native balsam firs in the Northeast but can be controlled with copper sprays, which will also aid in preventing needle-cast diseases. Avoid bark and branch injuries that induce cankers. Rust is taken care of, if necessary, by eliminating the proper alternate host, but only a specialist can identify the many different rust species.

FIRECRACKER PLANT (*Crossandra infundibuliformis*)

VIRUS. **Alternanthera Mosaic**, FL, MD, PA.

FIRETHORN

See Pyracantha.

FORESTIERA (Swamp-Privet)

MISTLETOE. **Phoradendron serotinum (flavescens)**, TX.

POWDERY MILDEW. **Microsphaera alni**, IL, TX.

ROT, Root. **Phymatotrichum omnivorum**, TX.

RUST. **Coleosporium minutum** (II, III), FL, TX; 0, I on pine; **Puccinia peridermiospora** (0, I), FL, TX; II, III on marsh grass.

FORSYTHIA (Goldenbells)

ANTHRACNOSE. **Gloeosporium** sp., OK.

BACTERIAL Grown Gall. **Agrobacterium tumefaciens**, MS, NJ, TX.

BLIGHT, Cane. **Botryosphaeria ribis**.

BLIGHT, Southern. **Sclerotium rolfsii**, GA.

BLIGHT, Twig. **Sclerotinia sclerotiorum**, NC; **Phytophthora nicotianae**, VA.

CANKER, Stem Gall. **Phomopsis** sp., KY.

LEAF SPOT. **Alternaria** sp., IA; **Phyllosticta discincola**, MD.

NEMATODE, Lesion. **Pratylenchus vulnus**.

NEMATODE, Root Knot. **Meloidogyne** sp., TX; **M. hapla**; **M. incognitaacrita**.

ROT, Root. **Phymatotrichum omnivorum**, TX.

VIRUS. **Tobacco Ring Spot,** MD.

FOUQUIERIA (Ocotillo, Candlewood)
ROT, Root. **Phymatotrichum omnivorum**, TX.
RUST. **Aecidium cannonii** (0, I), AZ.

FREMONTIA (Flannel Bush)
LEAF SPOT. **Ascochyta fremontiae**, CA.
ROT; Stem Girdle. **Phytophthora cactorum**, CA.
ROT, Crown, Root. **Phytophthora nicotianae** var. **nicotianae**, CA.
WILT. **Verticillium albo-atrum**, CA.

FRINGE-TREE (*Chionanthus*)
ANTHRACNOSE. **Discola fraxinea**, DE.
LEAF SPOT. **Cercospora chionanthi**, NJ to NC, WV; **Phyllosticta chionanthi**, NJ, WV; **Septoria chionanthi** and **S. eleospora**, SC, TX.
POWDERY MILDEW. **Phyllactinia corylea**, MD.
ROT, Root. **Phymatotrichum omnivorum**, TX.
ROT, Wood. **Daedalea confragosa**, MD.

FUCHSIA
BLIGHT, Gray Mold. **Botrytis cinerea**, AK, WV.
DIEBACK. **Phomopsis** sp., VA.
LEAF SPOT. **Septoria** sp., OK.
NEMATODE, Root Knot. **Meloidogyne** sp.
ROT, Mushroom Root. **Armillaria mellea**, CA.
ROT, Root. **Pythium rostratum**; **P. ultimum**, CA; **Phytophthora parasitica**, OR; **Thielaviopsis** sp., OR.
RUST. **Pucciniastrum epilobii**, NC, OR; **Uredo fuchsiae** (II), OH.
VIRUS. **Tomato Spotted Wilt**, CA; **Cucumber Mosaic**, MN.
WILT. **Verticillium albo-atrum**, CA.

Verticillium wilt is common in garden plantings of fuchsia in California.

G

GARDENIA (Cape-Jasmine)

Bacterial Leaf Spot. **Xanthomonas maculifolium-gardeniae**, CA.

Blight, Bud Rot. **Botrytis cinerea**, in greenhouses; outdoors in CA.

Canker, Stem Gall. **Phomopsis** (*Diaporthe*) **gardeniae**, CA, widespread in greenhouses, outdoors in CA, FL.

Leaf Spot. **Myrothecium roridum**, PA; **Pestalotia langloisii**, AL, FL; **Phyllosticta** sp., MS, NJ, TX; **Rhizoctonia** sp., NJ; **Mycosphaerella gardeniae**, GA.

Leaf Spot, Algal. **Cephaleuros virescens**, Gulf states.

Nematode, Burrowing. **Radopholus similis**, FL.

Nematode, Reniform. **Rotylenchulus reniformis**.

Nematode, Root Knot. **Meloidogyne hapla; M. incognita; M. javanica**; general; **M. arenaria**, NC.

Nematode, Spiral. **Helicotylenchus nannus**.

Nonparasitic. **Bud Drop**. Excessive soil moisture; temperature fluctuation.

Chlorosis. Soil too alkaline or soil temperature too low.

Dieback. Yellow veinbanding, defoliation from root smothering by overwatering.

Powdery Mildew. **Erysiphe polygoni**, TX.

Rot, Root. **Phymatotrichum omnivorum**, TX.

Sooty Mold. **Capnodium** spp., Gulf states.

In the deep South gardenia foliage is disfigured with sooty mold growing in whitefiy honeydew. Gardenias are difficult house plants, dropping buds with uneven humidity and temperature. To control Phomopsis canker take cuttings near top of plants. Substitute the more resistant Veitchii variety for susceptible Belmont and Hadley. Avoid syringing, which spreads bacterial leaf spot.

GARRYA (Tassel-Tree, Silk-Tassel Bush)

Leaf Spot. **Cercospora garryae**, CA, TX; **Phyllosticta garryae**, CA, TX; **Dothichiza garryae**, CA.

Rot, Root. **Phymatotrichum omnivorum**, TX.

Sooty Mold. **Lembosia lucens**, CA.

R.K. Horst, *Field Manual of Diseases on Trees and Shrubs*,
DOI 10.1007/978-94-007-5980-0_10,

GAULTHERIA (Checkerberry, Teaberry) (Source of Oil of Wintergreen)
BLOTCH, Sooty. **Gloeodes pomigena**, WI.
FRUIT SPOT; Black Speck. **Leptothyrella pomi**, WI; **Schizothyrium gaultheriae**, ME to VA, WI.
LEAF SPOT. **Cercospora gaultheriae**, NJ, WI; **Discohainesia oenotherae**, VA; **Mycosphaerella gaultheriae** (*Phyllosticta gaultheriae*), general; **Venturia arctostaphyli**, MD, MA, NJ, NY, VA.
POWDERY MILDEW. **Microsphaera alni**, MD.

GENISTA (Wood-waxen)
DIEBACK. **Diplodia** sp., CA, NJ.
LEAF SPOT. **Pleiochaeta setosa**, OR.
POWDERY MILDEW. **Erysiphe polygoni**, MO.
RUST. **Uromyces genistae-tinctoriae** (II, III), CA.
RUST, Soybean. **Phakopsora pachyrhizi**, Southern U.S.

GERMANDER (*Teucrium*)
DOWNY MILDEW. **Peronospora** sp., OK.
LEAF SPOT. **Cercospora teucrii**, NY to MS, TX, WI; **Phyllosticta decid-ua**, TX, WI.
NEMATODE, Root Knot. **Meloidogyne hapla**.
POWDERY MILDEW. **Golovinomyces cichoracearum**, IL, PA, WI.
RUST. **Puccinia menthae** (0, I, II, III), PA.

GILIA (Skyrocket)
BACTERIAL, MLO. **California Aster Yellows**, CA.
DOWNY MILDEW. **Peronospora giliae**, TX.
LEAF SPOT. **Ramularia giliae**, OR.
NEMATODE, Lesion. **Pratylenchus pratensis**, TX.
NEMATODE, Root Knot. **Meloidogyne** sp., CA.
NEMATODE, Stem. **Ditylenchus dipsaci**.
POWDERY MILDEW. **Sphaerotheca macularis**, CA, TX, WA.
ROT, Root. **Phymatotrichum omnivorum**, TX.
RUST. **Puccinia aristidae** (0, I), AZ, CO; **P. giliae** (II, III), AZ, CA, CO, NE, OR, WA; 0, I, unknown; **P. plumbaria** (0, I, III), CA, CO, NE, UT, WY; **P. yosemitana** (0, III), CA, CO. **Gymnosporangium exterum** (0, I), IN, KY, MO, NC, TN, VA, III on red-cedar.

GINKGO (Maidenhair-Tree)
ANTHRACNOSE. **Glomerella cingulata**, MD, TX.
BACTERIAL LEAF SPOT, Canker Stem. **Pseudomonas syringae** pv. **syringae**, CA.
LEAF SPOT. **Phyllosticta ginkgo**, PA; **Epicoccum purpurascens**, IL.
NEMATODE, Root Knot. **Meloidogyne** sp., MS.
ROT, Root. **Phymatotrichum omnivorum**, TX.
ROT, Sapwood. **Polyporus hirsutus**; **P. lacteus**; **P. tulipiferae**; **P. versicolor**; **Fomes connatus**, MD.
ROT, Seed. **Xylaria longeana**, PA.

GLOBE-MALLOW (*Sphaeralcea*)
Powdery Mildew. **Erysiphe polygoni**, ID.
Rust. **Puccinia interveniens**, ID, WA; **P. schedonnardi**, AZ, NM; **P. sherardiana**, AZ, CA, CO, ID, NE, NM, TX, UT, WY; **P. sphaeralceae** (I, III), CA.

GLORY-BUSH (*Tibouchina*)
Rot, Mushroom Root. **Clitocybe tabescens**, FL.

GOLDEN-CHAIN (*Laburnum*)
Blight, Twig. **Gibberella baccata**, NJ; **Fusarium** sp., MD, OH.
Canker, Stem. **Diplodia mutila**, OH.
Leaf Spot. **Cercospora laburni**, OK.

GOLDEN-LARCH (*Pseudolarix*)
Canker. **Dasyscyphus willkommii**, MA.

GOLDENRAIN-TREE (*Koelreuteria*)
Canker. **Nectria cinnabarina**, CA, CT.
Leaf Spot. **Cercospora** sp., FL.
Rot, Root. **Phymatotrichum omnivorum**, TX.
Wilt. **Verticillium** sp., NJ.

GOOSEBERRY (*Ribes*)
Anthracnose. **Pseudopeziza ribis**, general.
Blight, Cane. **Botryosphaeria ribis**, NJ, VA; **Leptosphaeria coniothyrium**, IN, MO.
Blight, Thread. **Pellicularia koleroga**, FL.
Canker; Dieback. **Nectria cinnabarina**, coral spot; **Botrytis cinerea**, CT, OR, WA; **Phragmodothella ribesia**, twig knot; **Physalospora obtusa**, NY to VA, KS.
Dodder, **Cuscuta** sp., MN, NY.
Downy Mildew. **Plasmopara ribicola**, MT, OR, WV, WI.
Leaf Spot. **Marssonina grossulariae**, OH, WI; **Cercospora angulata**, NY to MI, MO; **Mycosphaerella ribis**, MA to VA, AL, KS, MN, OR; **Phyllosticta grossulariae**, CT, NJ, WA, WI; **Ramularia** sp., MI.
Nematode, Leaf, Bud. **Aphelenchoides ritzemabosi**, CA.
Nonparasitic. **Leaf Blotch**. Magnesium deficiency, OR.
Leaf Scorch. Potassium deficiency.
Powdery Mildew. **Sphaerotheca macularis**, MN; **S. mors-uvae**, general; **Microsphaera grossulariae**, CA; **Phyllactinia corylea**, NY.
Rot, Root. **Armillaria mellea**, OR, WA; **Dematophora** sp.; **Xylaria hypoxylon**, OR; **Phymatotrichum omnivorum**, TX.
Rust. **Puccinia caricina** var. **grossulariata** (0, I) leaves, fruit, ME to MD, AK, ID, KS, MS, II, III on *Carex*; **P. caricina** var. **uniporula**, IA, MD, NY, WI; **Coleosporium jonesii** (II, III), CO, MO, MN, NM, WI, WY; **Cronartium occidentale**, AZ, CO, UT; **C. ribicola**, white pine blister rust (II, III) – European varieties are resistant.

Spot Anthracnose, Scab. **Sphaceloma ribis**, WA.
Virus. **Mosaic**. Unidentified, IL, NY.

Powdery mildew caused by *Sphaerotheca mors-uvae* is probably the most important gooseberry disease but is readily controlled by a lime-sulfur spray immediately after bloom, followed by bordeaux mixture spray for leaf spots.

GORDONIA (Franklinia and Loblolly-Bay)
Black Mildew, **Meliola cryptocarpa**, FL, LA.
Leaf Spot. **Phyllosticta gordoniae**, FL.
Rot, Root. **Phymatotrichum omnivorum**, TX.

GRAPEFRUIT
See Citrus Diseases.

GRASS-of-PARNASSUS (*Parnassia*)
Powdery Mildew. **Erysiphe polygoni**, NY.
Rust. **Puccinia parnassiae** (III), UT; **P. caricina** var. **uliginosa**, (I), AK; II, III on *Carex*.

GREVILLEA (Silk-Oak)
Dieback; Gum Disease. **Diplodia** sp. (*Physalospora rhodina*), FL.
Leaf Spot, Algal. **Cephaleuros virescens**, FL.
Nematode, Root Knot. **Meloidogyne** sp., CA.
Rot, Root. **Phymatotrichum omnivorum**, AZ.

GRINDELIA (Gumweed)
Leaf Spot. **Cercospora grindeliae**, CA, TX, WI; **Septoria grindeliae**, CO, KS, TX; **S. grindeliicola**, WI.
Powdery Mildew. **Golovinomyces cichoracearum**, MN to NM, CA, MT.
Rot, Root. **Phymatotrichum omnivorum**, TX.
Rust. **Coleosporium asterum** (II, III), CA, CO, WA, WI; 0, I on pine; **Puccinia dioicae** (0, I), KS, NE, TX; II, III on *Carex*; **P. grindeliae** (III), NE to TX, CA, MT; **P. stipae** (0, I), CO, KS, NE, ND, SD; II, III on *Stipa* and other grasses; **Uromyces junci** (0, I); II, III on *Juncus*.
Smut, Inflorescence. **Thecaphora californica**, CA, UT; **T. Cuneata**, CO, KS, NE, NM.

GROUNDNUT (*Apios*)
Leaf and Stem Disease. **Cercospora zebrina** (clover isolate), NC.
Virus. **Cucumber Mosaic**, LA; **Desmodium Yellow Mottle**, LA.

GUAVA (*Psidium*)
Anthracnose; Leaf and ruit Spot; Ripe Rot. **Glomerella cingulata**, FL.
Blight, Thread. **Pellicularia koleroga**, FL.
Leaf Spot. **Cercospora psidii**, FL.
Leaf Spot, Algal; Green Scurf. **Cephaleuros virescens**, FL.
Nematode, Burrowing. **Radopholus similis**, FL.
Nematode, Root Knot. **Meloidogyne** sp., FL.

Rot, Fruit. **Alternaria citri**, CA; **Rhizopus stolonifer**, HI.
Rot, Root. **Clitocybe tabescens**, FL; **Phymatotrichum omnivorum**, TX.
Rot, Wound. **Polyporus versicolor**, FL.
Scab. **Pestalotiopsis** spp., HI.

GUAYULE (*Parthenium*)
Rot, Root. **Pythium aphanidermatum**, AZ.

H

HACKBERRY, SUGARBERRY (*Celtis*)

BLIGHT, Leaf. **Cylindrosporium defoliatum**, TX.

BLIGHT, Thread. **Pellicularia koleroga**, FL.

CANKER, Felt Fungus. **Septobasidium burtii**, TX; **S. sydowii**, TX.

DOWNY MILDEW. **Pseudoperonospora celtidis**, GA, MD.

LEAF SPOT. **Cercosporella celtidis**, AL, TX, central states; **Cylindrosporium celtidis**, AL; **Phloeospora celtidis**, MA to TX; **Phyllosticta celtidis**, general; **Septogloeum celtidis**, NY.

MISTLETOE. **Phoradendron tomentosum** and **P. serotinum (flavescens)**, southeastern and Gulf states.

POWDERY MILDEW. **Sphaerotheca phytoptophila**, associated with gall mites causing witches' broom, widespread, especially in central states; **Uncinula macrospora**, FL, GA; **U. parvula**, widespread; **U. polychaeta**, in South.

ROT, Root. **Armillaria mellea**, occasional; **Helicobasidium purpureum**, seedling blight, TX; **Phymatotrichum omnivorum**, TX; **Poria ambigua**, TX.

ROT, Wood. **Daedalea** spp.; **Fomes** spp.; **Ganoderma lucidum**; **Polyporus** spp.; **Stereum bicolor**, cosmopolitan.

VIRUS. **Island Chlorosis**, IL; **Hackberry Leaf Mosaic; Angular Yellow Leaf Spot,** MN.

Hackberry witches' broom, so prominent in a winter landscape, are caused by gall mites in association with powdery mildew. There is no real control, except to cut off unsightly branches. Chinese hackberry and southern hackberry are less susceptible.

HALESIA (Silver-Bell, Snowdrop-Tree)

LEAF SPOT. **Cercospora halesiae**, TN; **Cristulariella moricola**, Zonate Leaf Spot, LA.

ROT, Wood. **Polyporus halesiae**, GA.

Other fungi are reported on twigs and branches but not as causing specific diseases.

R.K. Horst, *Field Manual of Diseases on Trees and Shrubs*,
DOI 10.1007/978-94-007-5980-0_11,

HAMELIA (Scarlet-Bush)
ROT, Root. **Clitocybe tabescens**, FL.
RUST. **Uredo hameliae**, FL.

HARDENBERGIA
NEMATODE, Root Knot. **Meloidogyne** sp.

HARDY ORANGE (*Poncirus*)
ANTHRACNOSE; DIEBACK. **Glomerella cingulata**, Gulf states.
CANKER; DIEBACK. **Diaporthe citri**, melanose, Gulf states; **Diplodianatalensis**, gummosis, general.
ROT, Root. **Phymatotrichum omnivorum**, TX; **Xylaria polymorpha**, LA.
SPOT ANTHRACNOSE; Citrus Scab. **Elsinoë fawcettii**, MS.

HAWTHORN (*Crataegus*)
BACTERIAL Fire Blight. **Erwinia amylovora**, widespread, especially on English hawthorn.
BLIGHT, Leaf. **Fabraea thuemenii** (*F. maculata, Entomosporium thuemenii*), East and central states, southward.
BLIGHT, Leaf; Fruit Spot. **Monilinia johnsonii**, NY to MN.
BLIGHT, Seedling. **Sclerotium rolfsii**, FL.
CANKER, Felt Fungus. **Septobasidium** spp., Southeast and Gulf states.
LEAF SPOT. **Cercospora apiifoliae**, TX; **Cercosporella mirabilis**, CO, NY, WI; **Coniothyrium pyrinum**, IL; **Cylindrosporium brevispina**, CA to MT, WA; **C. crataegi**, WA; **Gloeosporium crataegi**, WA; **Hendersonia crataegicola**, AL, TX; **Monochaetia crataegi**, FL; **Phyllosticta** spp., widespread; **Septoria crataegi**, MI to ND; **Mycosphaerella** sp., GA.
MISTLETOE. **Phoradendron serotinum (flavescens)**, TX; **Viscum album**, CA.
POWDERY MILDEW. **Podosphaera oxyacanthae**, East and central states; **Phyllactinia corylea**, general.
ROT, Gray Mold, on fruit. **Botrytis cinerea**, AK.
ROT, Root. **Armillaria mellea**, OK; **Phymatotrichum omnivorum**, TX. **Xylaria digitata**; **X. hypoxylon**, IN, OH.
ROT, Wood. **Daedalea confragosa**, cosmopolitan; **Polyporus versicolor**, cosmopolitan; **Fomes** spp.
RUST. **Gymnosporangium bethelii** (0, I), on fruits; ND to WA, CO, NM; III on juniper; **G. clavipes**, quince rust (0, I), on leaves, fruits, general, east of Rocky Mts.; **G. exiguum**, on leaves, fruits, TX; II, III on juniper; **G. floriforme**, on leaves, SC to FL, OK, TX; III on red-cedar; **G. globosum**, hawthorn rust (0, I) on leaves, ME to ND, FL, TX; II, III on juniper; **G. hyalinum** (0, I), leaves, Atlantic Coast, NC to FL; III unknown; **G. libocedri** (0, I), leaves and fruit, Pacific Coast; III on *Libocedrus*; **G. nelsonii** (0, I), on leaves, fruits, WY; III on juniper; **G. trachysorum** (0, I), on leaves, Atlantic and Gulf coasts; III on red-cedar; **G. tubulatum** (0, I), on leaves, Rocky Mts. to OR and WA.
SCAB. **Venturia inaequalis**, widespread; **V. crataegi**, FL.

Rusts are common on hawthorn and particularly so in the central states, where junipers should be kept at a distance. Leaf blight, beginning in a moist spring, causes spotting and defoliation in August. When branches die back from fire blight, cut them off several inches below the visible blighted portion.

HAZELNUT, FILBERT (*Corylus*)

BACTERIAL Blight. **Xanthomonas corylina**, OR, WA; **Pseudomonas syringae** pv. **coryli**.

BACTERIAL Crown Gall. **Agrobacterium tumefaciens**, WA.

BACTERIAL Spot. **Pseudomonas columnae**, IL.

BLIGHT, Eastern Filbert. **Anisogramma anomala**, NJ, NY, OR, WA.

CANKER; Twig Blight. **Apioporthe anomala**, eastern and central states; **Hymenochaete agglutinans**, ME; **Gleosporium** sp.

FILBERT STUNT. Undetermined causal agent, OR.

KERNEL SPOT; Yeast Spot. **Nematospora coryli**, OR.

LEAF BLISTER. **Taphrina coryli**, CT, MA, WI.

LEAF SPOT. **Cercospora corylina**, OK; **Cylindrosporium vermiforme**, WI; **Gloeosporium coryli**, ME to NJ, OR, PA, WA; **Phyllosticta coryli**, widespread; **Septogloeum profusum**, IN, MA, MS; **Septoria corylina**, MA to NJ, OR, WA, WI; **Sphaerognomonia carpinea**, GA; **Gnomoniella coryli**, general; **Alternaria arborescens**, **A. alternata**, **A. tenuissima** (leaf spot and brown/gray apical necrosis), AZ.

NONPARASITIC. **Bitter Pit** and **Brown Stain** of nuts. Cause unknown, OR, WA. **Shrivel**, "Blanks." Defective pollination, OR, WA.

POWDERY MILDEW. **Phyllactinia corylea**, northeastern and north central states, OR, WA; **Microsphaera alni**, northeastern and north central states; **P. gutata**, WA.

ROT, Root. **Armillaria mellea**, CA, OR, WA; **Phymatotrichum omnivorum**, TX.

ROT, Wood. **Poria ferrea**, CA.

VIRUS. **Apple Mosaic**, OR; **Clover Yellow Edge Phytoplasma**, OR.

Bacterial blight often kills young trees in Washington and Oregon. Powdery mildew is important in the Northwest.

HEATH (*Erica*)

POWDERY MILDEW. **Erysiphe polygoni**, CA.

ROT, Collar. **Phytophthora cinnamomi**, CA, NY.

ROT, Root and Crown. **Cylindrocladium pauciramosum**, CA.

ROT, Stem. **Ascochyta** sp., NJ.

RUST. **Pucciniastrum ericae** (II), CA.

HEATHER (*Calluna*)

NONPARASITIC. **Chlorosis**. Iron deficiency, corrected by spraying with ferrous sulfate.

ROT, Root, Collar. **Phytophthora cinnamomi**, CA, OR.

HEBE

BACTERIAL Fasciation. **Clavibacter fascians**, CA.

Leaf Spot. **Septoria exotica**, CA.
Rot, Root. **Armillaria mellea**, CA.
Wilt. **Fusarium oxysporum** f. sp. **hebae**, CA.

HEMLOCK (*Tsuga*)
Blight, Needle. **Didymascella tsugae**, MA, NH, WI; **Rosellinia herpotrichioides**, NC.
Blight, Shoot Tip. **Sirococcus tsugae**, GA.
Blight, Snow. **Herpotrichia nigra**, MT to OR, AK.
Blight, Twig. **Botrytis cinerea**, ID.
Canker, Bleeding. **Phacidiopycnis pseudotsugae** (*Phacidiella coniferarum*) on eastern hemlock.
Canker, Branch. **Discocainia treleasi** on western hemlock, AK, WA.
Canker, Stem Girdle, of saplings. **Hymenochaete agglutinans**, PA.
Canker, Twig. **Cytospora** sp., MD, VA; **Dermatea balsamea**, GA, NY, TN, VA.
Damping-Off. **Rhizoctonia solani**, cosmopolitan.
Mistletoe, Dwarf. **Arceuthobium campylopodium**, AK, ID; **A. cyanocarpum**, OR; **A. tsugense**, AK; **A. laricis**, WA; **A. abietinum**, OR.
Nonparasitic. **Canker;** Bark Split. Heavy soil and poor drainage.
Rot, Root. **Armillaria mellea**, MA to PA, MI, OR, WA; **Phymatotrichum omnivorum**, TX; **Fusarium solani**, PA; **F. oxysporum**, PA; **F. avenaceum**, PA; **Phytophthora citricola**, NC.
Rot, Seedling Root. **Cylindrocladium scoparium**, NY; **Rhizina undulata**, NY.
Rot, Wood. **Fomes** spp.; **Ganoderma lucidum**, Northeast; **Polyporus** spp., general.
Rust, Needle, Cone. **Caeoma dubium** (0, I), ID, MT, OR, WA; **Melampsora abietis-canadensis** (0, I), New England to PA; II, III on poplar; **M. farlowii** (III), New England to NC, WI; **Pucciniastrum hydrangeae** (0, I), IN, MD, PA, TN; II, III on hydrangea; **P. myrtilli** (0, I), ME to AL, IN, WI; II, III on *Ericaceae*; **Uraecium holwayi** (0, I), AK, MT, OR, WA; **Thekopsora minima**.

HERCULES-CLUB (*Aralia spinosa*)
Canker; Dieback. **Botryosphaeria ribis**, GA, VA, WV; **Nectria cinnabarina**, VA, WV.
Leaf Spot. **Cercospora atromaculans**, LA, TX; **Phyllosticta araliae**, TX; **P. everhartii**, TX, WV; **Stagonospora** sp., WV.
Rot, Root. **Phymatotrichum omnivorum**, TX; **Rhizoctonia solani**, TX.
Rot, Sapwood. **Polyporus tulipiferae**, MD.
Spot Anthracnose. **Sphaceloma araliae**, MD, MO.

HERCULES-CLUB (*Zanthoxylum clava-herculis*)
Leaf Spot. **Cercospora xanthoxyli**, FL, GA, TX; **Septoria pachyspora**, TX.
Mistletoe. **Phoradendron serotinum** (**flavescens**), TX.
Rust. **Puccinia andropogonis** var. **xanthoxyli** (0, I) FL, TX: II, III on *Andropogon*.

HIBBERTIA (Guinea-Gold-Vine)
Virus. **Cucumber Mosaic**, CA, SC.

HIBISCUS (Arborescent Forms) (Rose-of-Sharon, Confederate-Rose, Chinese Hibiscus, Kenaf)

BACTERIAL Crown Gall. **Agrobacterium tumefaciens**, MS.

BACTERIAL Leaf Spot. **Pseudomonas syringae**, CA, FL; **P. syringae** pv. **hibisci**, FL; **Xanthomonas campestris** pv. **malvacearum**, FL.

BACTERIAL Wilt. **Xanthomonas solanacearum**, FL.

BLIGHT, Blossom. **Choanephora infundibulifera**, FL.

BLIGHT, Leaf Stem. **Phytophthora palmivora**, LA; **P. parasitica**, FL.

BLIGHT, Thread. **Pellicularia koleroga**, FL; web, **P. filamentosa**, FL.

CANKER; DIEBACK. **Colletotrichum hibisci**, FL, TX.

LEAF SPOT. **Alternaria tenuis**, IN, NJ, PA; **Cercospora hibisci**, FL, OK, TX; **C. amalayensis**, GA; **Phyllosticta hibiscina**, LA; **Cristulariella pyramidalis**, FL, GA, MD, MS; **C. moricola**, LA.

NEMATODE, Leaf. **Aphelenchoides fragariae**, FL.

NEMATODE, Root Knot. **Meloidogyne** spp., MS, TX.

NONPARASITIC. **Strapleaf**. Molybdenum deficiency, FL.

POWDERY MILDEW. **Leveillula taurica**, TX.

ROT, Root. **Clitocybe tabescens**, FL; **Phymatotrichum omnivorum**, TX.

ROT, Stem. **Fusarium** sp., FL; **Phytophthora parasitica**, TX; **P. cactorum**, LA; **P. palmivora**, LA; **Sclerotinia sclerotiorum**, NY.

RUST. **Kuehneola malvicola** (**II, III**), Gulf states.

STUNTING. **Hoplolaimus magnistylus**, MS.

VIRUS. **Mosaic**, unidentified, FL; **Hibiscus Chlorotic Ringspot; Tomato Spotted Wilt**, TX; **Potato Leaf Roll**, Pacific Northwest; **Hibiscus Latent Fort Pierce**, NM.

HICKORY (*Carya*)

ANTHRACNOSE; Leaf Spot. **Gnomonia caryae**, general (see under Leaf Spots).

BACTERIAL Crown Gall. **Agrobacterium tumefaciens**, KS, MD, TX.

BLOTCH, Leaf. **Mycosphaerella dendroides**, widespread East and South.

CANKER. **Nectria galligena** and **Ceratocystis smalleyi**, eastern U.S.; **Strumella coryneoidea**, MD, PA.

CANKER; DECLINE, Bitternut. **Fusarium solani** and **Ceratocystis smalleyi**, IA, MN, WI.

CANKER, Bark Patch. **Aleurodiscus candidus**, MO, OH, PA; **Solenia ochracea**, MA.

CANKER, Felt Fungus. **Septobasidium** spp., Southeast.

CANKER, Heart Rot. **Poria spiculosa**, destructive to pignut, shagbark, PA, and South.

CANKER, Twig. **Rosellinia caryae**, MI; **Botryosphaeria berengeriana**.

GALL, Trunk, Branch. **Phomopsis** sp., eastern states.

LEAF SPOT. **Fusarium carpineum**, WI; **Hendersonia davisii**, WI; **Marssonina juglandis** (*Gnomonia leptostyla*), NJ, NC to IA; **Microstroma juglandis**, Witches' Broom; **Monochaetia desmazierii**, MD to NC, TN; **Mycosphaerella carigena**, WI; **Pestalotia sphaerelloides**, LA; **Phyllosticta caryae**, widespread; **Septoria caryae**, DE, MI; **S. hicoriae**, TX; **Cercospora fusca**.

Mistletoe. **Phoradendron serotinum (flavescens)**, IN, TX; **Viscumalbum**, CA.
Nematode, Root Knot. **Meloidogyne** sp.
Nonparasitic. **Rosette**. Zinc deficiency, southern states.
Powdery Mildew. **Phyllactinia corylea**, MI, WI; **Microsphaera alni**, cosmopolitan.
Rot, Heart. **Favolus alveolaris**, NY, VT, VA; **Fomes** spp., cosmopolitan.
Rot, Root. **Armillaria mellea**, cosmopolitan; **Clitocybe parasitica**, OK; **Phymatotrichum omnivorum** AZ, TX; **Helicobasidium purpureum**.
Rot, Wood. **Daedalea** spp.; **Ganoderma curtisii**; **Polyporus** spp.; **Steccherinum septentrionale**; **Schizophyllum commune**; **Stereum hirsutum**; **Trametes rigida**.
Scab; Leaf Spot. **Cladosporium effusum**, general.
Spot Anthracnose. **Elsinoë randii**.
Wilt. **Verticillium** sp., VA.

Of the various leaf spots on hickory, anthracnose caused by *Gnomonia caryae* is the most destructive. There are no recommended control measures for witches' brooms appearing on shagbark hickory except cutting them out.

HOLLY (*Ilex*) (*Ilex* opaca, American; *I. equifolium*, English; *I. cornuta*, Chinese; *I. crenata*, Japanese)
Bacterial Blight. **Clavibacter ilicis**, MA.
Black Mildew. **Asterina** (*Englerulaster*) **orbicularis**, DE to FL, TX; **A. ilicis**; **A. pelliculosa**; **Morenoella** (*Lembosia*) **ilicis**, Gulf states.
Blight, Flower. **Botrytis cinerea**, NJ.
Blight, Leaf and Twig. **Phytophthora ilicis**, OR, WA.
Blight, Thread. **Pellicularia koleroga**, LA; silky thread, **Rhizoctonia ramicola**; web, **Cylindrocladium scoparium**, AL.
Canker; Dieback. **Boydia insculpta**, secondary, OR, WA; **Diaporthe eres**, secondary; **Diplodia** sp., MD; **Nectria coccinea**; **Phomopsis** sp.; **Physalospora ilicis** on twigs, also leaf spot, CA, NJ, NY, NC, SC, TX, WA, WV; **Gloeosporium** sp., AL.
Canker, Felt Fungus. **Septobasidium** spp., FL, LA, NC.
Dodder. **Cuscuta compacta**, FL; **C. exaltata**, TX.
Leaf Spot. **Cercospora ilicis, NJ** and Gulf states; **Gloeosporium aquifolli**, NJ, TX, WA; **Macrophoma phacediella**, NJ; **Microthyriella cuticulosa**, black spot; **Phyllosticta concomitans**, LA; **P. terminalis**, MS; **Septoria ilicifolia**, NJ; **Cuticularia ilicis**, OR; **Sclerophoma** sp., OR.
Leaf Spot. **Sclerophoma** sp., Kentucky on English Holly (*Ilex aquifolium*).
Leaf Spot, Tar. **Phacidium curtisii**, MA to FL, TX; **Rhytisma ilicinicolum**, VA to GA, TN; **R. velatum**, GA, MS, SC.
Leaf Spot; Twig Dieback. **Cylindrocladium avesiculatum**, GA.
Nematode, Lance. **Hoplolaimus uniformis**, RI.
Nematode, Root Knot. **Meloidogyne incognita**; **M. incognita**; **M. arenaria**, NC; **M. hapla**; **M. javanica**.
Nonparasitic. **Leaf Blotch**. Purplish blotches associated with soil deficiencies and drought or winter injury.

Chlorosis. Mineral deficiency from excess of lime.
Spine Injury; **Copper Spray Injury**; **Boron Deficiency**.
POWDERY MILDEW. **Microsphaera alni**, AL, FL, IL, NC; **Phyllactinia corylea**, TN.
ROT, Leaf. **Rhizoctonia solani**, of cuttings.
ROT, Root. **Clitocybe tabescens**, FL; **Phymatotrichum omnivorum**, TX; **Corticium galactinum**, white root, MD.
ROT, Wood. **Daldinia vernicosa**, MD; **Daedalea confragosa**, MD; **Polyporus** spp.; **Stereum** spp., MD; **Ustulina deusta**, MD; **Poria** spp.; **Pleurotus ostreatus**, MD.
RUST. **Chrysomyxa ilicina** (II, III), NC, TN, WV, 0, I unknown.
SOOTY MOLD. **Capnodium elongatum**, Gulf states; **Fumago vagans**, OR, WA.
SPOT ANTHRACNOSE. **Elsinoë ilicis**, PA, on Chinese holly.

None of these diseases is as important to the average gardener as the holly leaf miner. Spraying is necessary for that insect but seldom obligatory for the various leaf spots. The most common leaf discoloration is a purplish blotch due to environment rather than a fungus. Proper growing conditions and correct soil acidity are important. Pick off occasional spotted leaves; prune out blighted branches.

HOLLY OSMANTHUS (*Osmanthus ilicifolius*)
LEAF SPOT. **Phyllosticta oleae**, TX.
NEMATODE, Root Knot. **Meloidogyne** sp.
ROT, Root. **Phymatotrichum omnivorum**, TX; **Rosellinia necatrix**, CA.
SOOTY MOLD. **Capnodium elongatum**, TX; **Fumago salicina**, TX.
WILT. **Verticillium albo-atrum**, VA.

HOLLY, WEST INDIAN (*Leea coccinea*)
LEAF SPOT, Blight, Defoliation. **Phytophthora meadii**, HI.

HOLODISCUS (Ocean Spray, Rock-Spirea)
CANKER, Twig; Coral Spot. **Nectria cinnabarina**, WA.
LEAF SPOT. **Cylindrosporium ariaefolium**, OR; **Rhopalidium cercosporelloidis**, ID, **Septogloeum schizonoti**, WA.
POWDERY MILDEW. **Podosphaera oxyacanthae**, ID, **Phyllactinia corylea**, WA.
VIRUS. **Holodiscus Witches' Broom**, OR, WA.

HONEY LOCUST (*Gleditsia*)
BACTERIAL Hairy Root. **Agrobacterium rhizogenes**, central states.
CANKER. **Dothiorella** sp., MS; **Cytospora gleditschiae**, IL; **Physalospora obtusa**.
CANKER, Felt Fungus. **Septobasidium curtisii**, AR, LA.
CANKER, Stem and Branch. **Kaskaskia gleditsiae**, IL; **Tubercularia ulmea**, ND; **Nectria cinnabarina**, MN; **Thyronectria austro-americana**, KS, NC.
CANKER; Wilt. **Thyronectria austro-americana**, CA, MA; **T. denigrata**, MA to SC, CA, LA, NE.
LEAF SPOT. **Cercospora condensata**, IL to KS, NE, WI.
LEAF SPOT, Tar. **Linospora gleditsiae**, SC, NC to NE, TX.
MISTLETOE. **Phoradendron serotinum (flavescens)**, IN, TX.
POWDERY MILDEW. **Microsphaera alni**, widespread.

Rot, Crown. **Phytophthora citrophthora**, CA.
Rot, Root. **Phymatotrichum omnivorum**, OK, TX; **Xylaria mali**, VA.
Rot, Wood. **Fomes** spp., widespread; **Daedalea ambigua**; **D. elegans**; **Ganoderma curtisii** and **G. lucidum**; **Polyporus** spp.; **Schizophyllum commune**.
Rust. **Ravenelia opaca** (III), IL.
Virus. **Robinia Brooming**, KY, OH.

HONEY PLANT, BITTERBUSH (*Picramnia*)
Spot, Tar. **Phyllachora domingensis**, FL.

HONEYSUCKLE (*Lonicera*)
Bacterial Crown Gall. **Agrobacterium tumefaciens**, CT.
Bacterial Hairy Root. **Agrobacterium rhizogenes**, central states.
Blight, Gray Mold. **Botrytis cinerea**, AK, CT.
Blight, Leaf. **Herpobasidium deformans**, CT, IA, NY, WI.
Blight, Sudden Oak Death. **Phytophthora ramorum**, CA.
Blight, Thread. **Pellicularia koleroga**, LA.
Blight, Twig. **Phoma mariae**, CT, MA, NY.
Leaf Spot. **Cercospora antipus**, MI to MT, TX; **C. varia**, TX; **Guignardia lonicerae**, CA; **Kabatia lonicerae**, CA; **Lasiobotrys affinis**, CA; **L. lonicerae**, Black Spot, OR; **Leptothyrium periclymeni**, widespread; **Marssonina lonicerae**, CA; **Mycosphaerella clymenia**, VA; **Septoria sambucina**, MT, WI; **S. xylostei**, WI.
Nematode, Root Knot. **Meloidogyne** sp.
Powdery Mildew. **Microsphaera alni**, general; **Erysiphe polygoni**, CA, WY.
Rot, Collar, Wood. **Fomes** spp.
Rot, Root. **Armillaria mellea**, NJ; **Phymatotrichum omnivorum**, TX.
Rust. **Puccinia festucae** (I), IL, IA.
Virus. **Mosaic**, MD; **Tobacco Streak**, MD.

HOP (*Humulus*)
Anthracnose; Leaf Spot. **Colletotrichum** sp.; **Glomerella cingulata**, IN, KS, MD, NY, OR, WA, WI.
Bacterial Crown Gall. **Agrobacterium tumefaciens**, CA, OK, OR, WA.
Blight, Cone Tip. **Fusarium sambucinum**, **F. avenaceum**, OR.
Downy Mildew. **Pseudoperonospora humuli**, CA, NY, OR, WA,WI, Argentina.
Leaf Spot. **Cercospora** sp., NE; **Cylindrosporium humuli**, NY to NC, IA, WI; **Phyllosticta decidua**, IA, WI; **P. humuli**, IA, MA, MI; **Mycosphaerella erysiphina**, CA; **Septoria humuli**, NH; **S. lupulina**, KS.
Nematode, Root Knot. **Meloidogyne** sp., CA.
Powdery Mildew. **Golovinomyces cichoracearum**; **Sphaerotheca macularis**, general.
Rot, Root. **Armillaria mellea**, OR.
Rust. **Aecidium** sp., WA.
Sooty Mold. **Fumago vagans**, CA, OR, WA.
Virus. **Hop Mosaic**, NY, OR, WA; **Hop Latent**; **Hop American Latent**, WA.
Wilt. **Verticillium albo-atrum**, ME, OH, OR, WI.

HOP-HORNBEAM, IRONWOOD (*Ostrya*)

CANKER. **Nectria** sp., NY; **Strumella coryneoidea**, MD, PA, WV.

CANKER, Bark Patch. **Aleurodiscus** spp.; Felt Fungus, **Septobasidium** spp.

LEAF BLISTER. **Taphrina virginica**, NH to FL, TX, WI.

LEAF SPOT. **Cylindrosporium dearnessi**, VA; **Gloeosporium robergei**, PA, WI; **Septoria ostryae**, IA, NY, WI.

POWDERY MILDEW. **Microsphaera alni**, widespread; **Phyllactinia corylea**, NY to FL, TX, WI; **Uncinula macrospora**, MI, WI.

ROT, Root. **Armillaria mellea**, cosmopolitan; **Clitocybe tabescens**, FL; **Phymatotrichum omnivorum**, TX.

ROT, Wood. **Daedalea confragosa**, cosmopolitan; **Fomes** spp.; **Pleurotus similis**, NY; **Poria** spp.; **Polyporus** spp.; **Stereum** spp.; **Trametes mollis**, NY, VT.

RUST. **Melampsoridium carpini** (II, III), NY; 0, I, unknown.

HOP-TREE (*Ptelea*)

LEAF SPOT. **Cercospora affiata**, IN, MO, TX; **C. pteleae**, TX; **Phloeospora pteleae**, TX; **Phyllosticta pteleicola**, IL; **Septoria pteleae**, WI.

ROT, Root. **Phymatotrichum omnivorum**, TX.

RUST. **Puccinia windsoriae** (0, I), NY to AL, KS; II, III on grasses.

HORNBEAM (*Carpinus*)

BLIGHT, Twig, Dieback. **Gibberella baccata**, AL.

CANKER, Bark Patch. **Aleurodiscus oakesii**, NY.

CANKER, Branch, Trunk. **Nectria galligena**, CT, NY; **Pezicula carpinea**, MA to GA, OK.

CANKER, Felt Fungus. **Septobasidium** spp., VA to FL, LA.

LEAF BLISTER. **Taphrina australis**, AL, CT, KY.

LEAF SPOT. **Clasterosporium cornigerum**, MD, NY, WI; **Cylindrosporium dearnessi**, MI; **Phyllosticta** sp., OK; **Sphaerognomonia carpinea**, GA, NY, PA, WV, WI;, **Cercoseptoria caryigena**, WI.

POWDERY MILDEW. **Microsphaera alni**, MA, IN, IA, MI, TX; **Phyllactinia corylea**, AL, IN, OH, TX, WI.

ROT, Root. **Armillaria mellea**, FL; **Phymatotrichum omnivorum**, TX.

ROT, Wood. **Daedalea confragosa**, NC; **Polyporus** spp., general; **Poria** spp., general; **Steccherinum** sp., CT; **Stereum** spp., general; **Daldinia concentrica**, MD; **Schizophyllum commune**, cosmopolitan.

HORSE-CHESTNUT, BUCKEYE (*Aesculus*)

ANTHRACNOSE; Leaf Blight. **Glomerella cingulata**, CT, MD, MO, NJ, NY, TX.

BLIGHT, Twig. **Botryosphaeria ribis** var. **chromogena**, MD, NY, GA.

BLOTCH, Leaf. **Guignardia aesculi** (*Phyllosticta paviae*), general.

CANKER; DIEBACK. **Nectria cinnabarina**, cosmopolitan; **Phytophthora cactorum**, bleeding canker, RI; **Gibberella acuminatum**, CA.

LEAF BLISTER, Yellow. **Taphrina aesculi**, CA, TX.

LEAF SPOT. **Cercospora aesculina** FL, VA; **Macrosporium baccatum**, KS; **Monochaetia desmazierii**, NC; **Micosphaerella maculiformis** var. **hippocastani**,

CA; **Septoria glabra**, IN; **S. hippocastani**, PA, VT, (Sudden Oak Death) **Phytophthora ramorum**, CA.

MISTLETOE. **Phoradendron serotinum** (**flavescens**), South, central states; **P. villosum**, CA, OR, TX; **Viscum album**, CA, on *Aesculus california* (California Buckeye).

NONPARASITIC. **Scorch**; **Leaf Scald**. Response to drought and heat.

POWDERY MILDEW. **Phyllactinia corylea**, CA, TX; **Uncinula flexuosa**, widespread.

ROT, Root. **Armillaria mellea**, cosmopolitan; **Phymatotrichum omnivorum**, TX.

ROT, Wood. **Collybia velutipes**, RI; **Ganoderma applanatum**, cosmopolitan; **Polyporus** spp.

RUST. **Aecidium aesculi**, IN, KS, MO, NE.

WILT. **Verticillium albo-atrum**, PA.

In a wet season leaves are blotched, turn brown, and drop from the *Guignardia* fungus; in a dry season leaves look scorched or blotched, turn brown, and drop from drought and heat. The minute black fruiting bodies of the fungus distinguish the parasitic form from the physiogenic disease.

HOYA

LEAF SPOT. **Myrothecium roridum**, FL.

HUCKLEBERRY (*Gaylussacia*)

BLACK MILDEW. **Dimerosporium ellissi**, MS, NJ.

DODDER. **Cuscuta** sp., PA.

GALL, Leaf. **Exobasidium vaccinii**, ME to VA, AL, FL, WI; **Synchytrium vaccinii**, red lead, NJ.

LEAF SPOT. **Ophiodothella vaccinii**, TX; **Pestalotia vaccinii**, WV; **Phyllosticta** sp., NJ, PA; **Ramularia effusa**, WI; **Rhytisma vaccinii**, tar spot, MA, OK, SC.

NEMATODE, Root Knot. **Meloidogyne** sp., OK.

POWDERY MILDEW. **Microsphaera alni** var. **vaccinii**, widespread.

RUST. **Pucciniastrum myrtilli** (II, III), ME to VA, WI; 0, I on hemlock.

HUCKLEBERRY (*Vaccinium*)

BLIGHT, Sudden Oak Death. **Phytophthora ramorum**, CA.

VIRUS. **Blueberry Scorch**. British Columbia, Canada.

HUISACHE, SWEET ACACIA (*Acacia farnesiana*)

ROT, Root. **Clitocybe tabescens**, FL; **Phymatotrichum omnivorum**, TX.

RUST. **Ravenelia australis**, TX; **R. hieronymi**, witches' broom, TX; **R. siliquae**, NM, TX.

HYDRANGEA

BACTERIAL LEAF SPOT. **Xanthomonas campestris**, GA.

BACTERIAL, MLO. **Hydrangea virescence**.

BACTERIAL Wilt. **Pseudomonas solanacearum**, NY.

BLIGHT, Flower; Gray Mold. **Botrytis cinerea**, cosmopolitan, especially after frost.

BLIGHT, Southern. **Sclerotium rolfsii**, FL.

LEAF SPOT. **Ascochyta hydrangeae**, AK, NJ; **Cercospora arborescentis**, IL, OK; **C. hydrangeae**, MD, VA to AL, FL, TX; **Colletotrichum** sp., MD, NJ; **Phyllosticta hydrangeae**, IN; **Septoria hydrangeae**, CT, MS, OH; **Corynespora cassiicola**, FL, GA; **Phoma exigua; Myrothecium roridum,** TN.

NEMATODE, Leaf. **Aphelenchoides** sp.

NEMATODE, Lesion. **Pratylenchus pratensis**, TX.

NEMATODE, Root Knot. **Meloidogyne** spp., MS, OK, TX.

NEMATODE, Stem, Bulb. **Ditylenchus dipsaci**.

NONPARASITIC. **Chlorosis**. Iron deficiency, excess lime.

POWDERY MILDEW. **Erysiphe polygoni**, general.

ROT, Root. **Phymatotrichum omnivorum**, TX; **Armillaria mellea**, CA.

ROT, Stem. **Rhizoctonia solani**, MD; wound, **Polyporus versicolor**, CT, IA, MD.

RUST. **Pucciniastrum hydrangeae** (II, III), PA to AR, IL; 0, I on hemlock.

VIRUS. **Hydrangea Ring Spot**; **Tomato Spotted Wilt**; **Hydrangea Latent; Chlorotic Mottle,** MN, NY.

HYSSOP (*Hyssopus*)

NEMATODE, Root Knot. **Meloidogyne** sp., MI.

I

ICE PLANT (*Carpobrotus*)
ROT, Root. **Pythium aphanidermatum**, CA; **Phytophthora cryptogea**, CA.
WILT. **Verticillium dahliae**, CA; **Phomopsis** sp., CA.

INCENSE-CEDAR (*Libocedrus* [= *Calocedrus decurrens*])
BACTERIAL Crown Gall. **Agrobacterium tumefaciens**, AZ, CA.
BLIGHT, Brown Felt. **Herpotrichia nigra**, CA.
CANKER, Branch. **Coryneum cardinale**, CA.
MISTLETOE. **Phoradendron juniperinum**, CA, NV, OR.
NEEDLE CAST. **Lophodermium juniperinum**, CA, OR.
ROT, Root. **Phymatotrichum omnivorum**, TX; **Armillaria mellea**, OR.
ROT, Wood. **Fomes pini** and **F. pinicola**, CA, OR; **Lenzites saepiaria**, general; **Polyporus** spp.; **Trametes isabellina**, CA.
RUST. **Gymnosporangium libocedri**, gall, witches' broom (II, III), CA, NV, OR; 0, I on apple and other *Malaceae*.

INDIA-HAWTHORN (*Raphiolepis*)
BACTERIAL, Fire Blight. **Erwinia amylovora**, FL, GA, LA.
BLIGHT, Southern. **Sclerotium rolfsii**, FL.
LEAF SPOT. **Entomosporium maculatum** (*Fabraea maculata*), CA.

INDIGO (*Indigofera*)
CANKER, STEM. **Diaporthe phaseolorum** var. **caulivora**, LA.
ROT, Root. **Phymatotrichum omnivorum**, TX.
RUST. **Ravenelia laevis**, TX; **Uromyces indigoferae** (II, III), FL, TX.

INDIGO-BUSH, LEAD-PLANT (*Amorpha*)
LEAF SPOT. **Cylindrosporium passaloroides**, widespread; **Diplodia amorphae**, TX.
POWDERY MILDEW. **Erysiphe polygoni**, IA, MN, WY.
ROT, Root. **Phymatotrichum omnivorum**, TX.
RUST. **Uropyxis amorphae** (0, I, II, III), general.

R.K. Horst, *Field Manual of Diseases on Trees and Shrubs*,
DOI 10.1007/978-94-007-5980-0_12,

INKBERRY (Ilex glabra)

BLACK MILDEW. **Asterina ilicis**, NJ; **Morenoella ilicis**, FL, GA, MS.

BLIGHT, Thread. **Pellicularia koleroga**, LA.

CANKER, Felt Fungus. **Septobasidium** spp., FL.

DODDER. **Cuscuta compacta**, FL.

LEAF SPOT. **Cercospora ilicis**, AL, FL, MS, NJ; **Phacidium curtisii** and **P. sphaerodieum**, tar spot, MA to FL, and TX; **Phytophthora cinnamomi**, (leaf spot, yellowing and wilting), VA.

ROT, Wood. **Poria versipora**, AR, LA.

IRONWEED (*Vernonia*)

BLACK MILDEW. **Stigmella vernoniae**, MO.

DOWNY MILDEW. **Plasmopara halstedii**, KS, MO.

LEAF SPOT. **Cercospora noveboracensis**, MO; **Ascochyta treleasei**, WI; **Mycosphaerella** sp.; **Septoria** sp., MO; **C. oculata**, WV to AL, NE, TX, WI.

POWDERY MILDEW. **Golovinomyces cichoracearum**, general; **Sphaerotheca macularis**, MO.

ROT, Root. **Phymatotrichum omnivorum**, TX.

RUST. **Coleosporium vernoniae** (I, III), general; 0, I on pine; **Puccinia vernoniae** (0, I, II, III), widespread.

IVY, BOSTON (*Parthenocissus tricuspidata*)

CANKER; DIEBACK. **Cladosporium herbarum**, IN, NJ, OH, PA.

DOWNY MILDEW. **Plasmopara viticola**, NJ, TX.

LEAF SPOT. **Cercospora amelopsidis**, MS; **C. pustulata**, TX; **Guignardia bidwellii** f. sp. **parthenocissi** widespread; **Phloeospora ampelopsidis**, IA; **Sphaeropsis hedericola**, NJ.

NEMATODE, Dagger. **Xiphinema index**, CA.

POWDERY MILDEW. **Uncinula necator**, TX.

ROT, Root. **Phymatotrichum omnivorum**, TX.

ROT, Stem. **Pellicularia filamentosa**, CT.

The Guignardia leaf spot is common and disfiguring.

IVY, ENGLISH (*Hedera helix*)

ANTHRACNOSE. **Amerosporium** (*Colletotrichum*) **trichellum**, MA to SC, OK, OR, TX, WA.

BACTERIAL SPOT. **Xanthomonas hederae**, GA, IL, MD, NJ, NY, VA, WA.

DODDER. **Cuscuta** sp., AZ, NJ.

LEAF SPOT. **Glomerella cingulata**, CT, MD, NY, TX; **Phyllosticta concentrica**, MA to AL, CA, NE, TX. WA; **Ramularia hedericola**, TX; **Sphaeropsis hedericola**, NY, WV; **Macrophoma** sp., TX.

MOLD, Leaf. **Cladosporium brunneolum**, CA.

NONPARASITIC. **Winter Injury**. Sunburn, Freezing.

POWDERY MILDEW. **Golovinomyces cichoracearum**, OK; **Oidium araliacearum**, CA.

Rot, Root. **Phytophthora palmivora**, CA; **Rhizoctonia solani**, CT; **Phymatotrichum omnivorum**, TX.

Sooty Mold. Common under trees, on insect exudate.

Spot Anthracnose; Ivy Scab. **Sphaceloma hederae**, CA, NC, VA.

Leaf spots are not often important in the garden but browning of foliage from winter injury is conspicuous in early spring. English ivy used as a ground cover is often black, with sooty mold growing in honeydew dropped from aphids in trees overhead. Dodder sometimes gets a strangle hold and is difficult to eradicate.

J

JACARANDA

ROT, Root. **Armillaria mellea**, CA; **Clitocybe tabescens**, FL; **Phymatotrichum omnivorum**, TX.

JACK-BEAN, SWORD BEAN (*Canavalia*)

LEAF SPOT. **Cercospora ternateae**, AL.

NEMATODE, Root Knot. **Meloidogyne** spp., FL. Usually resistant.

POD SPOT. **Vermicularia capsici**, FL; **V. polytricha**, AL.

VIRUS. **Potato Virus X**, WI.

JACOBINIA

NEMATODE, Burrowing. **Radopholus similis**, FL.

JACQUEMONTIA (Small Flower Morning Glory)

CANKER, STEM. **Diaporthe phaseolorum** var. **caulivora**, LA.

LEAF SPOT. **Cercospora alabamensis**, MS.

NEMATODE, Reniform. **Rotylenchulus reniformis**, GA, LA.

NEMATODE, Root Knot. **Meloidogyne** sp., AL; **Meloidogyne incognita**, LA.

ROT, Root. **Streptomyces ipomoea**, LA; **Ceratocystis fimbriata**, LA; **Fusarium oxysporum** f. sp. **batatas**, LA; **Plenodomus destruens**, LA; **Monilochaetes infuscans**, LA.

RUST. **Coleosporium ipomoeae** (II, III), LA.

VIRUS. **Tomato Spotted Wilt**, GA, TX.

WHITE RUST. **Albugo ipomoeae-panduratae**, AL.

JACQUINIA

BLACK SPOT. **Asterella paupercula**, FL.

JAMESIA (Cliffbrush)

BLIGHT. **Ovbularia edwiniae**, CO.

JASMINE (*Jasminum*)

BACTERIAL Crown Gall. **Agrobacterium tumefaciens**, MD.

BLIGHT, Blossom. **Choanephora infundibulifera**, FL.

R.K. Horst, *Field Manual of Diseases on Trees and Shrubs*,
DOI 10.1007/978-94-007-5980-0_13,

Blight, Southern. **Sclerotium rolfsii**, FL.
Canker, Stem Gall. **Phomopsis** sp., FL; **Phoma** sp., TX.
Leaf Spot. **Colletotrichum gloeosporioides**, FL, TX.
Leaf Spot, Algal; Green Scurf. **Cephaleuros virescens**, FL.
Nematode, Root Knot. **Meloidogyne** sp., FL.
Rot, Root. **Clitocybe tabescens**, FL; **Corticium galactinum**, MD.
Spot Anthracnose. **Elsinoë jasminae**, FL.
Virus. **Tobacco Ring Spot,** MD; **Variegation; Infectious Chlorosis**.

JATROPHA
Black Mildew. **Meliola jatrophae**.
Leaf Spot. **Colletotrichum gloeosporioides**, FL.
Necrotic Spot, Brown. **Cercospora jatrophae-curcas**, Brazil.
Rot, Root. **Clitocybe tabescens**, FL; **Phymatotrichum omnivorum**, TX.
Rust. **Phakopsora jatrophicola** (II, III), TX; **P. arthuriana**, Brazil.

JETBEAD (*Rhodotypos*)
Anthracnose. **Gloeosporium** sp., IL.
Blight, Twig; Coral Spot. **Nectria cinnabarina**, MA.
Leaf Spot. **Ascochyta rhodotypi**, IL.

JOINTWEED (*Polygonello articulata*)
Rust. **Puccinia polygoni-amphibii** (II, III), FL, WI.

JOJOBA (*Simmondsia*)
Blight, Leaf. **Phytophthora parasitica**, AZ.

JUJUBE (*Zizyphus*)
Leaf Spot. **Cercospora jujubae**, FL.
Rust. **Phakopsora zizyphi-vulgaris**, FL.

JUNIPER, RED-CEDAR (*Juniperus*)
Bacterial Crown Gall. **Agrobacterium tumefaciens**, FL, MS.
Black Mildew. **Apiosporium pinophilum**, OR; **Dimerium juniperi**, CA; **Asterina cupressina**.
Blight, Brown Felt. **Herpotrichia nigra**, northern Rockies to Pacific Northwest.
Blight, Leaf. **Chloroscypha juniperina**, IA; **Pestalotia funerea**, secondary; **Cercospora sequoiae** var. **juniperi**, CT, IA, KY, MO, NE, OK, WI; **Exosporium glomerulosum**, NC, SC; **Stigmina juniperina**.
Blight, Nursery, **Phomopsis juniperovora**, MA to FL, KS, MN.
Blight, Tip Dieback. **Sclerophoma pythiophila**, WI.
Blight, Twig. **Phomopsis juniperovora**, VA.
Canker. **Botryosphaeria stevensii**, IA; **Coryneum cardinale**, CA; **Caliciopsis nigra**, gall, NY.
Canker, Bark Patch. **Aleurodiscus nivosus**, AL, OR, TX.
Dieback, Branch. **Diplodia mutila**, PA.
Mistletoe. **Phoradendron densum** and **P. juniperinum**, widespread.
Needle Cast; Leaf Spot. **Lophodermium juniperinum**, widespread, secondary; **Cylindrocarpon** sp., OR; **Kriegeria** sp., OR.

Nematode, Lesion. **Pratylenchus vulnus**, NC.
Parasite, "False Foxglove". **Aureolaria grandiflora** var. **serrata**, TX.
Phytoplasma, Witches' Broom, OR.
Rot, Heart. **Fomes juniperinus**, PA to KY, TN; **F. earlei**, Southwest; **F. subroseus**, general; **F. texanus**, AZ, NM, TX.
Rot, Root. **Armillariella mellea**, CA.
Rot, Root. **Clitocybe tabescens**, FL; **Phymatotrichum omnivorum**, OK, TX; **Phytophthora cactorum**; **P. citrophthora**, CA; **P. dreschleri**, NC; **P. lateralis**, NC; **P. syringae**, NC.
Rot, Root. **Fomes annosus**, VA.
Rot, Root and Crown. **Phytophthora cinnamomi**, CA; **P. cryptogea**, CA.
Rot, Wood. **Lenzites saepiaria**, occasional; **Coniophora corrugis**, Pacific Northwest; **Daedalea juniperina**, SC to AR.
Rust. **Gymnosporangium bermudianum** (III), gall on stems, Gulf states; **G. bethelii** (III), on stems, ND to OK, NM, WA; 0, I on hawthorn; **G. clavariiforme** (III), gall on stems, ME to AL west to MT; 0, I on *Amelanchier*, chokeberry, quince, pear; **G. clavipes** (III), quince rust, on stems, ME to IL, MT; 0, I on *Amelanchier*, hawthorn, quince, apple; **G. corniculans** (III), gall on stems, ME, MI, NY, WI; 0, I on *Amelanchier*; **G. cornutum** (III), gall on stems, leaves, CO, ME, MI; 0, I on mountain-ash; **G. davisii** (III), leaf gall, ME, WI; 0, I on chokeberry; **G. effusum** (III), gall on stems, NY to SC; 0, I on chokeberry; **G. exiguum** (III), on leaves, CA, OK, TX; 0, I on hawthorn; **G. exterum** (III), gall on stems, KY; 0, I, *Gillenia*; **G. floriforme** (III) gall, on leaves, stems, SC to FL, OK, TX; 0, I on hawthorn; **G. globosum**, hawthorn rust (III); 0, I on hawthorn, apple, pear, mountain-ash; **G. harknessianum** (III), on western juniper; 0, I on *Amelanchier*; **G. inconspicum** (III), CA, CO, UT; 0, I on *Amelanchier*; **G. gracile**, Witches' Broom (III), 0, I, on sand pear; **G. clavipes**, galls on stems, VA; **G. globosum**, gall on stems, VA; **G. exterum**, gall on stems, VA; **G. juvenescens** (III), gall on stems, Witches' Broom, MN, NE, WI; 0, I on *Amelanchier*; **G. japonicum** (III), CT, MA, NJ, WA; 0, I on *Photinia*; **G. kernianum**, Witches' Broom, ID, OR to AZ, NM; 0, I on *Amelanchier*, pear; **G. multiporum** (III), CO to CA, NM; 0, I unknown; **G. nelsonii** (III), gall on stems, MT, SD; 0, I on hawthorn, crabapple, mountain-ash, quince, *Amelanchier*; **G. nidusavis** (III), gall on stems; Witches' Broom, East and South; 0, I on *Amelanchier*, apple, hawthorn, mountain-ash, quince; **G. speciosum** (III), on stems, AZ, CO, NV, NM; 0, I on *Fendlera*, *Philadelphus*; **G. trachysorum** (III), on stems, FL, LA, MS, SC; **G. tremelloides** (III) gall on stems, CO to Pacific Northwest; 0, I on mountain-ash; **G. tubulatum** (III), gall on stems, SD to OR, WA; 0, I on hawthorn; **G. vauqueliniae** (III), on one-seed juniper; 0, I on *Vauquelinia*, causing Witches' Broom.
Rust, Cedar-Apple. **Gymnosporangium juniperi-virginianae** (III), gall on leaves of red-cedar, prostrate and Rocky Mountain junipers, general; 0, I on apple, crabapple.

Of this lengthy list of juniper rusts the three common apple rusts are most important in the garden, not so much for the damage to this host as for the harm to the deciduous fruit or ornamental. These three are the cedar-apple rust caused by *Gymnosporangium juniperi-virginianae*, the hawthorn rust by *G. globosum*, and the quince rust by *G. clavipes*. The latter is perennial in juniper and may produce spores each spring for as long as 20 years. Varieties of *Juniperus chinensis* and *J. communis* and some other forms are resistant to cedar-apple and cedar-hawthorn rusts. If you have only a few red-cedars, it is quite feasible to cut out galls in late winter, before spores are produced.

K

KAGENECKIA

Scab. **Spilocaea botryae**, CA.

KENTUCKY COFFEE-TREE (*Gymnocladus*)

Leaf Spot. **Cercospora gymnocladi**, North central states; **Marssonin** sp., Northeast; **Phyllosticta gymnocladi**, IL.

Rot, Root. **Phymatotrichum omnivorum**, OK, TX.

Rot, Wood. **Polyporus pulchellus**, IN, MI.

KERRIA

Blight, Twig. **Phomopsis japonica**, NJ, OH, TX.

Canker, Coral Spot. **Nectria cinnabarina**, NJ, OR, WA.

Leaf Spot; Leaf and Twig Blight. **Coccomyces kerriae**, widespread, eastern states to IA and TX; **Septoria** sp., MD, NJ.

Rot, Root. **Phymatotrichum omnivorum**, TX.

KIWI (*Actinidia*)

Bacterial Canker. **Pseudomonas syringae**, CA.

Nematode, Root Knot. **Meloidogyne incognita**, SC.

Rot, Root and Wilt. **Cylindrocladium crotalariae**, SC; **Pythium ultimum**, AL.

Wilt. **Verticillium albo-atrum**, Chile.

KUDZU (*Pueraria*)

Bacterial Blight. **Pseudomonas phaseolicola**, CT to FL, LA, IN; **P. syringae**, NY.

Blight, Web. **Pellicularia filamentosa**, GA, MS.

Damping-Off. **Rhizoctonia solani.**

Leaf Spot. **Alternaria** sp., secondary; **Mycosphaerella pueraricola**, MS, AL, GA.

Nematode, Root Knot. **Meloidogyne** spp., general.

Rot, Charcoal. **Macrophomina phaseoli**, GA; stem, **Fusarium** sp., LA.

Rot, Root. **Phymatotrichum omnivorum**, TX.

Rust. **Phakopsora pachyrhizi**, FL, IL, KY, NC, TX.

Virus. **Bean Common, Mottling** and **Ringspot,** MS.

R.K. Horst, *Field Manual of Diseases on Trees and Shrubs*,
DOI 10.1007/978-94-007-5980-0_14,

KUMQUAT (*Fortunella*)

DODDER. **Cuscuta campestris**, FL.

LEAF SPOT. **Cephaleuros virescens**, algal spot, Gulf states; **Phyllosticta citricola**, MS.

NEMATODE, Citrus. **Tylenchulus semipenetrans**, FL.

ROT, Black. **Alternaria citri**; stem-end, **Diaporthe citri**, CA.

L

LANTANA

Black Mildew. **Meliola cookeana**, FL.

Leaf Spot. **Alternaria** sp., TX.

Nematode, Leaf. **Aphelenchoides fragariae**, NJ.

Nematode, Root Knot. **Meloidogyne** spp., widespread.

Rot, Root. **Phymatotrichum omnivorum**, TX.

Rust. **Puccinia lantanae** (III), FL.

Virus. **Tomato Spotted Wilt**, PA.

Wilt. **Fusarium** sp., NJ.

LARCH (*Larix*)

Blight, Needle. **Hypodermella laricis**, Great Lakes, Pacific Northwest. **Meria laricis**, ID.

Blight, Seedling. **Thelephora caryophyllea**, ID, MT, girdle, smother; **Rhizina undulata**, ID; **Meria laricis**, WA.

Canker. **Aleurodiscus amorphus**, bark patch, Northwest; **A. spinulosus**; **A. werii**; **Lachnellula willkommii**, European larch canker, MA, eradicated, ME; **Valsa kunzei** var. **kunzei**.

Canker. **Phomopsis pseudotsugae** (*Phacidiella coniferarum*) on western larch, Northwest.

Damping-Off. **Rhizoctonia solani**, cosmopolitan.

Mistletoe, Dwarf. **Arceuthobium campylopodum**, MT to OR, WA; **A. pusillum**, Northeast, MN.

Needle Cast. **Lophodermium** sp., WA; **L. laricinum**, MT to OR; **L. laricis**, ID; **Cladosporium** sp., ME; **Mycosphaerella laricina**, IA, VT, WI.

Rot, Heart. **Echinodontium tinctorium**, WA; **Fomes** spp., widespread.

Rot, Root. **Armillaria mellea**, cosmopolitan; **Phymatotrichum omnivorum**, TX.

Rot, Sapwood. **Lenzites saepiaria**.

Rot, Seedling. **Botrytis douglasii**, gray mold, Northwest; **Cylindrocladium scoparium**, NJ; **Phytophthora cinnamomi**, MD; **Sparassis radicata**, MT to OR, WA.

R.K. Horst, *Field Manual of Diseases on Trees and Shrubs*,

DOI 10.1007/978-94-007-5980-0_15,

Rot, Wood. **Polyporus** spp.; **Poria** spp.; **Stereum** spp., widespread.
Rust. **Melampsora paradoxa** (0, I), northern U.S. including AK; II, III on willow; **M. larici-populina** (0, I), Quebec, Canada, WA; II, III on poplar; **M. medusae** (0, I), New England to MI and IN; II, III on poplar; **M. medusae** f. sp. **deltoidea** (0, I), WA; II, III on poplar; **Melampsoridium betulinum** (0, I), CT, WI; II, III on birch.
Wood Decay fungi associated with woodpecker nest cavities: **Coniophora puteana**, **Gloeophyllum sepiarium**, **Oligoporus placentus**, **Stereum sanguinolentum**, **Neolentinus lepideus**, **Phaelous schweinitzii**, **Wolfiporia cocos**, OR.

LAUREL
See California-Laurel, Cherry-Laurel, Mountain-Laurel.

LAUREL, SWEET BAY (*Laurus*)
Blight, Ramorum. **Phytophthora ramorum**, PA.
Blight, Thread. **Pellicularia koleroga**, SC.

LAURESTINUS (*Viburnum tinus*)
Downy Mildew. **Plasmopara viburni**, GA.
Leaf Spot. **Hendersonia tini**, LA; **Leptosphaeria tini**, LA.
Leaf Spot, Algal. **Cephaleuros virescens**, FL, LA.
Nematode, Root Knot. **Meloidogyne** sp., CA.
Powdery Mildew. **Erysiphe viburni**, CA.
Virus. **Alfalfa Mosaic**, Chile.
Wilt. **Verticillium albo-atrum**, OR.

LAVENDER (*Lavandula*)
Bacterial Leaf Disease. **Xanthomonas campestris**, CA.
Leaf Spot. **Septoria lavandulae**, OH, OK.
Nematode, Root Knot. **Meloidogyne** sp.
Rot, Root. **Armillaria mellea**, TX.

LEADTREE (*Leucaena*)
Rot, Root. **Ganoderma sulcatum**, TX.
Rust. **Ravenelia leucaenae** (II, III), TX.

LEATHERWOOD (*Dirca*)
Rust. **Puccinia dioicae** (I), ME to MN, MO, AL; II, III on *Carex*.

LEBBEK (*Albizzia lebbek*)
Leaf Spot, Algal. **Cephaleuros virescens**, FL.
Rust. **Sphaerophragmium acaciae**, FL.

LEDUM (Labrador-Tea)
Gall, Leaf. **Exobasidium vaccinii**, AK, OR, WA; **Synchytrium vaccinii**, red spot, ME.
Leaf Spot. **Ascochyta ledi**, WI; **Cryptostictis arbuti**, CA, OR; **Rhytisma andromedae**, tar spot, ID.
Powdery Mildew. **Microsphaera alni**, WA.

Rust. **Chrysomyxa ledi** (II, III), CA, CT, ID, MI, MT, NV, NH, NY, WI, WY; **C. ledicola** (II, III), AK, ME, NH, NY, WA, WI; 0, I on spruce.
Spot Anthracnose. **Elsinoë ledi**, CA, ME, MI, MN, NY, OR, PA, WA, WI.

LEMON
See Citrus Fruits.

LEPTOSPERMUM
Rot, Root. **Armillaria mellea**, CA.

LESPEDEZA
Leaf and Stem Disease. **Cercospora zebrina** (clover isolate), NC.

LEUCOSPERMUM
Blight. **Calonectria colhounii**, HI.
Wilt. **Verticillium dahliae**, CA.

LEUCOTHOË
Black Mildew. **Asterina diplodioides**, AL.
Blight, Leaf Spot. **Cylindrocladium avesiculatum**, GA; **C. colhounii**, NC.
Blight, Ramorum. **Phytophthora ramorum**, PA.
Canker, Felt Fungus. **Septobasidium pseudopedicellatum**, FL.
Gall, Leaf. **Exobasidium vaccinii**, MA, MS, NC.
Leaf Spot. **Cercospora leucothoës**, NJ, NY; **Cryptostictis** sp.; **Guignardia leucothoës**, MD, NC, RI, SC, TN, VA; **Mycosphaerella leucothoës**, NJ; **Phyllosticta terminalis**, NY to FL; **Ramularia andromedae**, NJ; **Rhytisma decolorans**, tar spot, CT, FL, TN, VA. **Pestalotia leucothoës**, NJ.
Powdery Mildew. **Microsphaera penicillata**, GA.
Spot Anthracnose. **Elsinoë ledi**, FL.

LIGUSTRUM
See Privet.

LILAC (*Syringa*)
Anthracnose, Shoot Blight. **Gloeosporium syringae**, CT, MA.
Bacterial Crown Gall. **Agrobacterium tumefaciens**, CT.
Bacterial Blight; Twig Canker. **Pseudomonas syringae**, northeastern states to AL, IL, Pacific Coast.
Blight, Blossom, Shoot. **Phytophthora cactorum**, IA, MD, MA, MN, NJ; **P. syringae**, MD, NY; **Sclerotinia sclerotiorum**, WA.
Blight, Cutting. **Cylindrocladium scoparium** and **C. floridanum**, FL, OH.
Blight, Gray Mold. **Botrytis cinerea**. Pacific Northwest, Northeast.
Blight, Ramorum. **Phytophthora ramorum**, PA.
Blight, Thread. **Pellicularia koleroga**, FL, MS, NC.
Bacterial, Witches Broom. **MLO**, IL.
Blotch, Leaf. **Heterosporium syringae**, NJ; **Cladosporium herbarum**, secondary, cosmopolitan.
Canker; Stem Girdle. **Hymenochaete agglutinans**, CT.

Canker; Dieback. **Physalospora obtusa**, MA to VA, OH.
Leaf Spot. **Cercospora lilacis**, widespread; **Macrophoma halstedii**, CT, NJ, NY; **Phyllosticta** sp., MA; **Pleospora herbarum**, secondary, MD.
Nematode, Citrus. **Tylenchulus semipenetrans**.
Nonparasitic. **Blight**. Graft incompatibility of lilac scion on privet stock.
Powdery Mildew. **Microsphaera alni**, general.
Phytoplasma. **Witches' Broom**, IL; **Ash Yellows**, MD, ND.
Rot, Root. **Phymatotrichum omnivorum**, TX; **Armillaria mellea**, CA, MS; **Thielaviopis basicola**, CT.
Rot, Wood. **Polyporus gilvus**, MD; **Stereum purpureum**, OK.
Virus. **Lilac Mottle**; **Lilac Ring Spot,** MI, MN; **Lilac Witches' Broom**, MD.

Powdery mildew is the most general and conspicuous disease of lilacs, but it comes too late in the season to damage the bushes materially. It can be controlled with repeated application of GreenCure, sulfur, or Karathane, where the time and expense are justified. In wet seasons bacterial and Phytophthora blights may be important, but dieback is more often due to borers than to diseases.

LINDEN, BASSWOOD (*Tilia*)
Anthracnose; Leaf Spot. **Gnomonia tiliae** (*Gloeosporium tiliae*), CT to VA, IA, MN.
Blight, Leaf. **Cercospora microsora**, general.
Canker, Bark. **Aleurodiscus acerinus**; **A. griseo-canus**, IA, MO, PA.
Canker; Dieback. **Botryosphaeria** sp., MD; **Nectria** spp., NY, PA, VA; **Strumella** sp., NJ.
Canker, Felt Fungus. **Septobasidium fumigatum**, FL.
Leaf Spot. **Phlyctema tiliae**, NY; **Phyllosticta praetervisa**, WI; **Sphaeropsis** sp., OK.
Mistletoe. **Phoradendron serotinum** (**flavescens**), South.
Powdery Mildew. **Microsphaera alni**, MN; **Phyllactinia corylea**, MN; **Uncinula clintonii**, general.
Rot, Heart. **Daedalea confragosa**, VT; **Fomes** spp.; **Steccherinum septentrionale**, AL, MI; **Pholiota adiposa**, MA, PA, TN.
Rot, Root. **Phymatotrichum omnivorum**, TX; **Ustulina vulgaris**, NY; **Phytophthora dreschleri**, NC.
Rot, Sapwood. **Collybia velutipes**, occasional; **Pleurotus ostreatus**, cosmopolitan.
Rot, Wood. **Daldinia concentrica**, MN, NY; **Lenzites betulina**, NY, VT; **Schizophyllum commune**, MN; **Polyporus** spp.; **Stereum** spp.; **Trametes mollis**, VT.
Sooty Mold. **Fumago vagans**.
Spot Anthracnose. **Elsinoë tiliae**, VA.
Wilt. **Verticillium albo-atrum**, IL.

Anthracnose and Cercospora leaf blight are common diseases.

LIPSTICK VINE (*Aeschynanthus*)
Leaf Spot. **Corynespora casiicola**, FL; **Myrothecium roridum**, FL.
Virus. **Tomato Spotted Wilt**, PA.

LITHOCARPUS (Tanbark Oak)
BLIGHT, Leaf. **Pestalotia castagnei**, CA.
BLIGHT, Sudden Oak Death. **Phytophthora ramorum**, CA. OR.
CANKER. **Phytophthora ramorum, P. nemorosa, P. gonapodyides, P. cambivora, P. siskiyouensis, P. taxon,** OR.
LEAF SPOT. **Ceuthocarpum conflictum**, CA.
POWDERY MILDEW. **Erysiphe trina**, CA.
ROT, Wood. **Poria** spp.; **Stereum hirsutum**.
RUST. **Cronartium quercuum** (II, III), CA; 0, I on pine.
WILT. **Phytophthora ramorum**, CA, OR.

LITSEA (Pond-Spice)
LEAF SPOT. **Cercospora olivacea**, GA.
WILT, Laurel. **Raffaelea lauricola**, FL.

LOCUST (*Robinia*)
ANTHRACNOSE; BLACK LOCUST. **Colletotrichum gloeosporioides**, GA, SC.
BACTERIAL, MLO. **Witches' Broom**, VA.
BLIGHT, Seedling; Leaf. **Alternaria** sp., NC to AL, MO; **Fusicladium robiniae**, MD to AL, MO, WI.
CANKER. **Nectria cinnabarina**, VA.
CANKER; Twig Blight. **Aglaospora anomala**, ME to GA; **Fusarium sarcochroum**, IA; **Diaporthe oncostoma**, NY to GA, IL.
DAMPING-OFF. **Rhizoctonia solani**, ME to AL, NE, TX; **Pythium** spp., NE, TX.
DODDER. **Cuscuta** sp.; **C. arvensis**, widespread.
LEAF SPOT. **Cladosporium epiphyllum**, TN, VA, WV; **Cylindrosporium solitarium**, TX; **Gloeosporium revolutum**, NJ; **Phloeospora robiniae**, NY to OH; **Phyllosticta robiniae**, LA.
MISTLETOE. **Phoradendron serotinum** (**flavescens**), AZ, NM, NC, TX; **Viscum album**, CA.
NEMATODE, Lesion. **Pratylenchus** sp., OR.
NEMATODE, Root Knot. **Meloidogyne** spp., OK, TX.
NONPARASITIC **Chlorosis**. Iron Deficiency, NE, TX.
Little Leaf. Zinc deficiency, CA.
POWDERY MILDEW. **Erysiphe polygoni**, CA; **Microsphaera diffusa**, IL, NC; **Phyllactinia corylea**, NM.
ROT, Heart. **Fomes** spp., **F. rimosus**, VA; **Polyporus** spp.
ROT, Root. **Armillaria mellea**; **Phymatotrichum omnivorum**, NM, OK; **Pythium myriotylum**, NC; **Fusarium** sp., AL, GA.
ROT, Seedling Stem. **Phytophthora cinnamomi**, MD; **Rhizoctonia bataticola**, AL, NC; **Sclerotium bataticola**, TX.
ROT, Wood. **Daedalea unicolor**, WI; **Poria** spp.
VIRUS. **Locust Witches' Broom**, Robinia Brooming Disease, PA to GA, AR, OH.
WILT, Seedling. **Phytophthora parasitica**, AL, NC, VA; **Verticillium albo-atrum**, IL.

LOGANBERRY
See Blackberry.

LOQUAT (Eriobotrya)
ANTHRACNOSE; Flower Blight; Withertip. **Colletotrichum gloeosporioides**, CA, FL, TX.
BACTERIAL Fire Blight. **Erwinia amylovora**, Gulf states, AZ, CA.
BLIGHT, Leaf; Blotch. **Fabraea maculata**, FL.
LEAF SPOT. **Phyllosticta eriobotryae**, FL; **Pestalotia** sp., secondary; **Septoria eriobotryae**, FL.
LEAF SPOT, Algal. **Cephaleuros virescens**, FL.
NEMATODE, Burrowing. **Radopholus similis**, FL.
ROT, Collar, Crown. **Phytophthora cactorum**, CA.
ROT, Root. **Armillaria mellea**, CA; **Clitocybe tabescens**, FL; **Phymatotrichum omnivorum**, TX.
SCAB. **Fusicladium eriobotryae**, on leaves, stems, fruit, Gulf states; **Spilocaea eriobotryae** (*Fusicladium photinicola*).

LUCUMA (Canistel, Egg-Fruit)
ANTHRACNOSE; Fruit Spot. **Colletotrichum gloeosporioides**, FL.
CANKER. **Physalospora obtusa**, SC.
RUST. **Achrotelium lucumae** (II, III), FL.

LYCHEE
ROT, Mushroom Rot. **Clitocybe tabescens**, sometimes fatal in FL.

LYCIUM (Desert-Thorn, Christmasberry)
LEAF SPOT. **Cercospora lycii**, OK.
POWDERY MILDEW. **Arthrocladiella mongeotii**, WA.
ROT, Root. **Phymatotrichum omnivorum**, TX.
RUST. **Aecidium lycii** (0, I), AZ; **Puccinia globosipes** (II, III), UT to NM, CA; 0, I unknown; **P. tumidipes** (II, III), TX to AZ, UT.

LYONIA (Maleberry, Fetterbush, Staggerbush)
BLACK MILDEW. **Asterina lepidigena**, FL; **Lembosia andromedae**, AL; **Morenoella dothideoides**, FL.
BLIGHT, Leaf. **Ramularia cylindriopsis**, ME.
DODDER. **Cuscuta compacta**, FL.
GALL, Leaf; Shoot Hypertrophy. **Exobasidium vaccinii**, MA to FL, AL.
LEAF SPOT. **Ceuthocarpon ferrugineum**, FL; **Pestalotia vaccinii**, secondary; **Septoria pulchella**, GA.
LEAF SPOT, Tar. **Rhytisma andromedae**, widespread; **R. decolorans**, MS.
POWDERY MILDEW. **Microsphaera alni** var. **vaccinii**, AL.
ROT, Wood. **Poria versipora**, ME.
RUST. **Pucciniastrum myrtilli** (II, III), DE to AL, AR; 0, I on hemlock.

LYSILOMA
ROT, White Pocket Heart. **Fomes extensus**, FL.
RUST. **Ravenelia annulata** (II, III), FL; **R. lysilomae**, FL.

M

MAACKIA

Rot, Root. **Phymatotrichum omnivorum**, TX.

MACADAMIA

Canker. **Phytophthora cinnamomi**, CA.

Decline, Quick. **Phytophthora tropicalis**, HI.

MADRONE (*Arbutus menziesii*)

Blight (Sudden Oak Death). **Phytophthora ramorum**, CA.

Canker. **Hendersonula toruloidea**, WA; **Fusicoccum** sp., western North America.

Canker, Trunk. **Phytophthora cactorum**, CA, WA.

Gall, Leaf; Red Leaf Spot. **Exobasidium vaccinii**, CA, OR, TX, WA.

Leaf Spot. **Ascochyta hanseni**, CA; **Cryptostictis arbuti**, CA, OR; **Didymosporium arbuticola**, OR; **Mycosphaerella arbuticola**, CA, OR, WA; **Phyllosticta fimbriata**, OR.

Leaf Spot, Tar. **Rhytisma arbuti**, CO, OR, TX, WA.

Rot, Heart. **Fomes subroseus**, OR.

Rot, Root. **Phymatotrichum omnivorum**, TX.

Rot, Wood. **Lenzites saepiaria**, OR, TX; **Polyporus** spp., OR, TX; **Trametes sepium**, CA.

Rust. **Puccinastrum sparsum** (II, III), CA, OR, WA.

Spot Anthracnose. **Elsinoë mattirolianum**, CA.

MAGNOLIA

Bacterial Leaf Spot. **Pseudomonas syringae**, IL; **P. cichorii**, AL.

Black Mildew. **Dimerosporium magnoliae**, TX; **Irene araliae**, MS; **Meliola amphitricha**, Gulf states; **M. magnoliae**; **Trichodothis comata**, South.

Blight, Seedling. **Rhizoctonia solani**, NJ.

Blight, Thread. **Pellicularia solani**, NJ.

Canker, Felt Fungus. **Septobasidium langoisii**, **S. tenui**, **S. leprieurii**, Gulf states.

DOI 10.1007/978-94-007-5980-0_16, © Springer Science+Business Media Dordrecht 2013

Canker; Twig Blight. **Nectria** sp., WV; **N. magnoliae**, NJ, TN, WV; **Tubercularia** sp., MI.

Leaf Spot. **Alternaria tenuis**, TX; **Cladosporium fasciculatum**, GA to TX; **Cercospora magnoliae** (*Mycosphaerella milleri*), FL, NJ, NC, SC, WV; **Cristulariella pyramidalis**; **Colletotrichum** sp., FL, GA, SC; **Coniothyrium** sp., TX; **C. olivaceum**, TX; **Epicoccum nigrum**, TX; **Exophoma magnoliae**, FL, TX; **Glomerella cingulata**, AL, FL, GA, LA, MS, SC, TX; **Hendersonia magnoliae**, VA; **Heterosporium magnoliae**, FL, NC, TX; **Micropeltis alabamensis**, AL; **cookei**, NY to Gulf states, CA; **P. magnoliae**, large leaf spot, NY to Gulf states, CA; **Septoria magnoliae**, SC to FL, TX.

Leaf Spot; Algal; Green Scurf; "Red Rust". **Cephaleuros virescens**, general.

Nematode, Burrowing. **Radopholus similis**, FL.

Nematode, Citrus. **Tylenchulus semipenetrans**, FL.

Nematode, Lance. **Hoplolaimus uniformis**, RI.

Nematode, Lesion. **Pratylenchus pratensis**.

Nematode, Root Knot. **Meloidogyne** sp.

Nematode, Sting. **Belonolaimus longicaudatus**, FL.

Parasitic Lichen. **Strigula elegans** and **S. complanata**, Southern U.S., LA.

Rot, Flower. **Ciborinia** (*Sclerotinia*) **gracilipes**, MD.

Rot, Heart. **Fomes fasciatus**; **F. geotropus**.

Rot, Root. **Phymatotrichum omnivorum**, TX; **Clitocybe tabescens**, GA.

Rot, Wood. **Daldinia concentrica**, cosmopolitan; **Polyporus** spp.; **Poria** spp., **Stereum** spp.

Spot Anthracnose. **Elsinoë magnoliae**, FL, GA, LA, MS.

Wilt. **Verticillium albo-atrum**, CA, IN.

The many leaf spots need not be alarming; they seldom cause premature defoliation. In the Gulf states lichens appear as small round gray spots on leaves, and the parasitic alga *Cephaleuros* often forms a velvety coating with hairlike outgrowths.

MAHOGANY (*Swietenia*)

Blight, Twig, Stem. **Phomopsis** sp., OR.

Leaf Spot. **Pestalotia swieteniae**, FL; **Phyllachora swieteniae**, FL.

MAHONIA (Oregon-Grape)

Canker; Leaf Blotch. **Leptosphaeria berberidis**, ID, MD.

Leaf Spot. **Cercospora** sp., LA; **Gloeosporium berberidis**, WA; **Phomopsis** sp., secondary; **Phyllosticta** spp., AL, CT, WA; **Cylindrocladium** sp., FL, GA.

Nematode, Root Knot. **Meloidogyne** sp.

Nonparasitic. **Scald**. In eastern states winter injury to foliage is severe.

Parasitic Lichen. **Strigula elegans** and **S. complanata**, Southern U.S., LA.

Rot, Root. **Phymatotrichum omnivorum**, TX.

Rust. **Cumminsiella mirabilissima** (0, I, II, III), common Western Great Plains to Pacific Coast; **C. texana**, TX; **C. wootoniana** (II, III), AZ, NM; **Puccinia graminis** (0, I), CA, MI, ND; II, III on cereals and grasses; **P. koeleriae**

(0, I), CO, ID, MT, OR; II, III on *Koeleria*; **P. oxalidis** (0, I), NM; II, III on *Oxalis*.

Rust due to *Cumminsiella* is usually inconspicuous, but in a wet season there is a general blighted effect. *Mahonia*, like barberry, is under wheat-rust quarantine. *Mahonia repens* is immune.

MALLOTUS
LEAF SPOT. **Cercospora malloti**, MS.

MALVAVISCUS
BLIGHT, Twig. **Sclerotinia sclerotiorum**, TX.
LEAF SPOT. **Phyllosticta malvavisci**, TX.
ROT, Root. **Clitocybe tabescens**, FL; **Helicobasidium purpureum**, TX; **Phymatotrichum omnivorum**, TX.
RUST. **Kuehneola malvicola** (II, III), TX; **Puccinia heterospora** (III), TX.

MANDEVILLA (*Mandevilla splendens*)
BACTERIAL Wilt. **Ralstonia solanacerum**, IN, MI, MN, OH.

MANGO (*Mangifera*)
ANTHRACNOSE; Flower and Twig Blight; Fruit Rot. **Glomerella cingulata**, cosmopolitan.
BLIGHT, Twig. **Phomopsis** sp., FL; **Physalospora** (*Diplodia*) spp.
CANKER, Brown Felt. **Septobasidium pilosum** and **S. pseudopedicellatum**, FL.
DIEBACK. **Neofusicoccum parvum**, Peru.
GALL, ON TREE. **Fusarium decemcellulare**, FL.
LEAF SPOT. **Phyllosticta mortoni**, FL, TX; **Septoria** sp., FL.
LEAF SPOT, Algal, Green Scurf. **Cephaleuros virescens**, general.
NONPARASITIC. **Little Leaf**. Zinc deficiency, FL.
Soft Nose. On Indian varieties.
POWDERY MILDEW. **Oidium mangiferae**, CA, FL.
ROT, Root. **Phymatotrichum omnivorum**, TX.
ROT, Root and Crown. **Phytophthora palmivora**, AZ.
ROT, Stem-end Fruit. **Diplodia** sp.
SOOTY MOLD. **Capnodium** spp., general.
SPOT ANTHRACNOSE; Mango Scab. **Elsinoë mangiferae**, FL.
WILT. **Verticillium albo-atrum**, FL.

MANGROVE (*Rhizophora*)
LEAF SPOT. **Cercospora rhizophorae**, FL.

MANIHOT (Cassava, Manioc)
ANTHRACNOSE; Withertip. **Gloeosporium manihotis**, FL, LA, TX.
DIEBACK. **Physalospora abdita** and **P. rhodina**, FL.
LEAF SPOT. **Cercospora henningsii**, FL, TX.
NEMATODE, Root Knot. **Meloidogyne** sp., AL, FL.

ROT, Root. **Phymatotrichum omnivorum**, TX; **Rhizoctonia solani**, FL.
VIRUS. **Cassava Common Mosaic**, FL.

MANZANITA (*Arctostaphylos*)
BLACK MILDEW. **Meliola** sp., CA.
BLIGHT (Sudden Oak Death). **Phytophthora ramorum**, CA.
GALL, Red Leaf Spot. **Exobasidium vaccinii**, widespread; shoot gall, **E. vaccinii-uliginosi**, CA, OR.
LEAF SPOT. **Cryptostictis arbuti**, OR; **Phyllosticta amicta**, CA, OR.
MISTLETOE. **Phoradendron villosum**, CA, OR.
ROT, Root. **Phymatotrichum omnivorum**, TX.
ROT, Root and Crown. **Phytophthora cinnamomi**, CA.
ROT, Wood. **Fomes annosus**, CA; **F. arctostaphyli**, and **F. igniarius**, general; **Poria ferruginosa**.
RUST. **Pucciniastrum sparsum** (II, III), CA, OR.

MAPLE (*Acer*)
ANTHRACNOSE; Leaf Blight. **Discula campestris**, WI; **Gloeosporium apocryptum**, large blotches, general; **G. acerinum**, small spots; **G. saccharinum**; **G. aceris**.
BACTERIAL Crown Gall. **Agrobacterium tumefaciens**, MI, OK, TX, VA.
BACTERIAL Leaf Scorch. **Xylella fastidiosa**, KY.
BACTERIAL Leaf Spot. **Pseudomonas aceris**, on Japanese maple, CA; **P. acernea**.
BLIGHT, Inflorescence. **Ciborinia acerina**, MA, NY.
BLIGHT, Leaf. **Didymosporina aceris**, CT, PA.
BLIGHT, Seedling; Smother. **Thelephora albido-brunnea**, NC, VA.
BLIGHT, Sudden Oak Death. **Phytophthora ramorum**, CA.
BLIGHT, Twig. **Macrophoma** sp., AR, NY.
CANKER, Bark. **Aleurodiscus acerinus**, MA, VT; **Hymenochaete agglutinans**, CT, MI, MT.
CANKER, Basal, Crown. **Phytophthora cinnamomi**, NJ; **P. cambivora**, NJ.
CANKER, Bleeding. **Phytophthora cactorum**, CT, MA, NJ, NV, NY, RI, VA; **P. citricola**, NV.
CANKER; DIEBACK. **Colletotrichum acutatum**, on Japanese maple, CT; **Coniothyrium negundinis**, IL, OK; **Cytospora** spp., MI; **Eutypella parasitica**, MI, MN, NH, NY, VT, WI; **Hypoxylon morsei**, MI, MN; **Fusarium** spp.; **Nectria cinnabarina**, cosmopolitan, often secondary; **N. coccinea**; **N. galligena**, widespread trunk canker; **Phomopsis** sp., OH; **P. acerina**, NY; **Physalospora** spp., secondary; **Sphaeropsis albescens**, IL, IA, KS, NY, ND, SD, WI; **Strumella coryneoidea**, MI, PA; **Schizoxylon microsporum**, Lake states; **Stegonsporium acerinum**, NJ.
CANKER, Felt Fungus. **Septobasidium** spp., AL, FL, KY, LA, NC, SC, TN, VA.
DAMPING-OFF. **Rhizoctonia solani**, cosmopolitan.
DODDER. **Cuscuta gronovii**, on seedlings, NY.
LEAF BLISTER. **Taphrina carveri**, AL, KY, MI, MO; **T. bartholomaei**, UT; **T. darkeri**, OR; **T. dearnessii**, black, GA, MI, MN, NC, NY, OK, PA, VA; **T.**

sacchari, on sugar maple, AR, GA, IN, KS, ME, MI, MO, NH, NY, OH, PA, TN, WV, WI.

LEAF SPOT. **Actinopelte dryina**, IL; **Alternaria** sp., secondary; **Cercospora negundinis**, KS, NE, WI; **C. saccharini**, MA; **Cercosporella aceris**, WA; **Cristulariella depraedens**, CT, NY; **C. pyramidalis**, FL; **Monochaetia desmazierii**, GA, NC, OK, TN; **Illosporium maculicola**, WA; **Laestadia brunnea**, NC, SC; **Leptothyrella acerinum**, OK; **Marssonina truncatula**, OR; **Pezizella oenotherae**, NY, NC, VA; **Phyllosticta minima**, eyespot, general; **P. negundinis**, ME to AL, TX, WI; **Piggotia negundinis**, leaf blotch, WI, WY; **Stilbella acerina**, PA; **Septoria aceris**, general; **Venturia acerina**, NY, PA, VA, WV, WI.

LEAF SPOT, Tar. **Rhytisma acerinum**, general; **R. punctatum**, speckled, general.

MISTLETOE. **Phoradendron serotinum (flavescens)**, common from NJ to FL, MO, TX; **Viscum album** on Big Leaf Maple (*Acer macrophyllum*) and Silver Maple (*A. saccharinum*), CA.

NEMATODE. **Criconemoides** sp.; **Hemicycliophora** sp.; **Pratylenchus sp. thornei**; **Tylenchorhynchus** sp.; **Xiphinema** sp.

NEMATODE, Root Knot. **Meloidogyne** sp., OR; **M. ovalis**, WI.

NONPARASITIC. **Leaf Scorch**. Common on street and lawn trees, associated often with high temperature after a moist spring.

Decline. Frequently roadside salt injury.

Frost Crack. Gas Injury.

PARASITIC LICHEN. **Strigula elegans** and **S. camplanata**, Southern U.S., LA.

POWDERY MILDEW. **Phyllactinia corylea**, CA, IA, NC, OR, SD, VT; **Uncinula circinata**, ME to AL, MI, MO, TX; **Microsphaera alni**; **Sawadaea bicornis**, ID, WA; **S. tulasnei**, NY, WI, Montreal, Canada.

ROT, Heart, Sapwood, Wound. **Collybia velutipes**, CT, MA; **Daedalea** spp., northeastern and north central states, southward; **Daldinia concentrica**, cosmopolitan; **Fomes applanatus**; **F. connatus**; **F. igniarius**; **Ganoderma lucidum**, fatal to some street trees, NJ, NY; **Hericium erinaceous**, VT to MD, MI, MN; **Lenzites** spp.; **Pholiota adiposa**; **Pleurotus** spp.; **Polyporus** spp.; **Poria** spp.; **Steccherinum septentrionale**, general; **Stereum** spp.; **Schizophyllum commune**, cosmopolitan; **Ustulina vulgaris**, northeastern and north central states; **Valsa leucosomoides**, on tapped sugar maples.

ROT, Root. **Armillaria mellea**, general; **Clitocybe tabescens**, GA, MO; **Helicobasidium purpureum**, TX; **Phymatotrichum omnivorum**, TX.

ROT, Sapstreak. **Endoconidiophora virescens**; **Ceratocystis coerulescens**.

ROT, Seedling, Charcoal. **Macrophomina phaseoli**, IL.

VIRUS. **Peach Rosette**; **Tobacco Necrotic Ring Spot**.

WILT. **Verticillium albo-atrum**, widespread in cultivated trees, especially Norway maple.

Verticillium wilt is the most destructive maple disease and is particularly prevalent in street trees. The wilting may be confined to a single branch, which can be cut out, or may kill the whole tree. In removing dead trees, get all of the root system and replace with a different kind of tree. Leaf scorch is common on sugar maple in hot, windy weather; anthracnose may be conspicuous in wet weather.

MATRIMONY-VINE (*Lycium halimifolium*)
LEAF SPOT. **Alternaria** sp., IA; **Cercospora lycii**, IA; **Phyllosticta lycii**, NY, OH.
POWDERY MILDEW. **Erysiphe polygoni**, CT, DE, MD, NJ, PA; **Microsphaera diffusa**, OH, PA, UT; **Sphaerotheca pannosa**, ID, WA.
RUST. **Puccinia tumidipes** (II, III), NY to AL, SD, TX; **P. globosipes** (II, III), PA.

MEDLAR (*Mespilus*)
BACTERIAL Fire Blight. **Erwinia amylovora**, NY.
LEAF SPOT. **Fabraea maculata**, CA.
RUST. **Gymnosporangium clavipes**, NY.

MELALEUCA
RUST. **Puccinia psidii**, CA, FL; **Uromyces striatus** (on sweetclover), KS.

MENTZELIA (Blazing Star)
LEAF SPOT. **Phyllosticta mentzeliae**, KS, TX; **Septoria mentzeliae**, KS, TX, WA.
ROT, Root. **Phymatotrichum omnivorum**, TX; **Rhizoctonia solani**, NJ.
RUST. **Puccinia aristidae** (0, I), AZ, CO; II, III on grasses; **Uredo floridana**, FL.

MENZIESIA
GALL, Leaf. **Exobasidium vaccinii**, Pacific Northwest, AL, NC, VA, WV.
LEAF SPOT, Tar. **Melasmia menziesii**, MT, and WY to AK, OR; **Rhytisma** sp.
POWDERY MILDEW. **Microsphaera alni** var. **vaccinii**, AK, VA.
RUST. **Pucciniastrum myrtilli** (II, III), WV; 0, I on hemlock.

MESQUITE (*Prosopis*)
BACTERIAL Crown Gall. **Agrobacterium tumefaciens**, TX.
BLIGHT, Leaf. **Cercospora prosopidis**; **Scleropycnium aureum**, AZ, TX.
LEAF SPOT. **Napicladium prosopodium**, TX; **Phyllosticta juliflora**, also pod spot, OK, TX; **Gloeosporium leguminum**, pod spot, TX.
MISTLETOE. **Phoradendron californicum**; **P. serotinum** (**flavescens**), TX to CA, and **P. tomentosum**, TX.
POWDERY MILDEW. **Leveillula taurica**; **Oidiopsis taurica**; **Uncinula prosopodis**, TX.
ROT, Heart. **Polyporus texanus**, CA, TX; **Fomes everhartii**, AZ, TX; **Schizophyllum commune**, TX; **Hypoxylon diatrypeoides** (dieback, trunk rot), Mexico.
ROT, Root. **Phymatotrichum omnivorum**, TX.
RUST. **Ravenelia arizonica** (II, III), TX to CA; **R. holwayi** (0, I, II, III), TX to CA.

MIMOSA, SILK-TREE (*Albizzia julibrissin*)
CANKER; DIEBACK. **Nectria cinnabarina**, DC, NC, VA.
NEMATODE, Root Knot. **Meloidogyne arenaria** var. **thamesii**; **M. incognita**; **M. javanica**; **M. hapla**.
NEMATODE, Stubby Root. **Trichodorus** primitivus.
ROT, Heart. **Ganoderma lucidum**; Root, **Armillaria mellea**.

Virus. **Mimosa Striped Chlorosis**, AR.

Wilt. **Fusarium oxysporum** f. sp. **perniciosum**, AL, AR, CA, GA, FL, MS, NJ, NC, SC, VA.

The mimosa wilt is one of the most devastating tree disease on record. The fungus is in the soil with no possibility of control by aerial spraying. The incidence of wilting is probably increased by nematodes. Resistant varieties Tryon and Charlotte have been released, but occasional specimens succumb to wilt. The fungus may also be seed-transmitted.

MOCK-ORANGE (*Philadelphus*)

Bacterial Blight. **Pseudomonas syringae**, MN.

Blight, Flower and Shoot. **Botrytis cinerea**, cosmopolitan.

Blotch, Sooty, **Sarcinella heterospora**, FL.

Leaf Spot. **Ascochyta philadelphi**, NY; **Cercospora angulata**, MO, TX; **Ramularia philadelphi**, TX, WA; **Septoria philadelphi**, ID, IA, MT.

Nematode, Root Knot. **Meloidogyne** sp.

Powdery Mildew. **Phyllactinia corylea**, GA, MT, WA.

Rot, Root. **Phymatotrichum omnivorum**, TX.

Rust. **Gymnosporangium speciosum** (0, I), CO, NM, UT; III on juniper.

MONKSHOOD VINE (*Ampelopsis aconitifolia*)

Dieback. **Tubercularia nigricans**, NH, TX.

Nematode, Dagger. **Xiphinema index**.

MOUNTAIN-ASH (*Sorbus*)

Bacterial Crown Gall. **Agrobacterium tumefaciens**, CT, NJ.

Bacterial Fire Blight. **Erwinia amylovora**, widespread, VA.

Blight, Leaf. **Fabraea maculata**, AK, MN, WV, WI.

Blight, Twig. **Nectria cinnabarina**, corla spot, AK, NC; **Phomopsis** sp., MA; **Valsa leucostoma**, OH, WV.

Canker. **Cytospora** sp., VA.

Canker, Blister. **Nummularia discreta**, IA, MA.

Canker, Branch; Fruit Rot. **Glomerella cingulata**, IN.

Canker; Dieback. **Cytospora chrysosperma**, MN, MT, NE, NJ, WA **C. leucostoma**, MT; **C. massariana**, ID; **C. microspora**, MT; **Fusicoccum** sp., IL.

Canker, Trunk; Black Rot. **Cytospora rubescens**, IA; **Physalospora obtusa**, IN, MI, OH, CT to VA.

Leaf Spot. **Alternaria** sp., IA; **Graphium sorbi**, NY, WI; **Phyllosticta globigera**, ID, WA; **P. sorbi**, IL, IA, ME, MO, OK, TX; **Septoria sorbi**, IA; **S. sitchensis**, ID.

Mistletoe, European. **Viscum album**, CA.

Powdery Mildew. **Podosphaera oxyacanthae** var. **tridactyla**, WA.

Rot, Heart. **Polyporus hirsutus**, MI, WA; **P. versicolor**, WI.

Rot, Root. **Armillaria mellea**, NJ; **Phymatotrichum omnivorum**, TX.

Rust. **Gymnosporangium cornutum** (0, I), ME to NJ, WI, MT to CO, AK, WA; III on juniper; **G. globosum** (0, I), OR; III on *Libocedrus*; **G. nelsonii** (0, I), MT,

WA, WY; III on juniper; **G. nootkatense** (0, I), AK, OR, WA; III on *Chamaecyparis*; **G. tremelloides** (0, I), MT to CO, AK, WA.
SCAB. **Venturia inaequalis**, IL, MN, NY, WA.

Mountain-ash is quite susceptible to fire blight, but affected branches can usually be pruned out. Rust may appear on foliage in midsummer. If the mountain-ash is more desirable than the junipers nearby, the latter can be eradicated. Fertilizing will help the tree recover from a bout with Cytospora canker but may increase susceptibility to fire blight.

MOUNTAIN-HEATHER (*Phyllodoce*)
BLIGHT, Brown Felt. **Herpotrichia nigra**, ID.
GALL, Leaf. **Exobasidium vaccinii-uliginosi**, OR.

MOUNTAIN-HOLLY (*Nemopanthus*)
LEAF SPOT. **Ramularia nemopanthus**, NY.
LEAF SPOT, Tar. **Rhytisma ilicis-canadensis**, ME to WV, MI.
POWDERY MILDEW. **Microsphaera alni**, NY, WI.
ROT, Wood. **Poria inermis**, PA.

MOUNTAIN-LAUREL (*Kalmia*)
BACTERIAL LEAF SPOT. **Pseudomonas syringae**, OR.
BLIGHT, Flower. **Ovulinia azaleae**, AL, SC, AL.
BLIGHT, Leaf. **Phomopsis kalmiae**, NY to NC.
BLIGHT, Necrotic Shoots and Leaves. **Phytophthora irrigate** and **P. hydropathica**, MD, PA, VA, WV.
CANKER; Felt Fungus. **Septobasidium** sp., AL, TX.
GALL, Leaf. **Exobasidium vaccinii**, AK; **Synchytrium vaccinii**, red spot, NJ.
HEART, Rot, Wood Rot. **Fomes annosus**, NC.
LEAF SPOT. **Cercospora kalmiae**, CT to AL, TN; **Mycosphaerella colorata** (*Phyllosticta kalmicola*), CT to AL, IN, MI, TX; **Pestalotia kalmicola**, secondary, DE, NJ, PA, TX, WA; **Septoria angustifolia**, MA to AL, OH; **Rhytisma andromedae**, tar spot, VT.
NONPARASITIC. **Chlorosis**. Usually iron deficiency.
POWDERY MILDEW. **Microsphaera alni** var. **vaccinii**, WI.
ROT, Root. **Armillaria mellea**, MD; **Corticium galactinum**, MD; **Phymatotrichum omnivorum**, TX; **Phytophthora lateralis**, NC; **P. cinnamomi**, VA; **P. palmivora**, VA.
ROT, Wood. **Polyporus versicolor**, VA; **Stereum rameale**, VA.
VIRUS. **Rhododendron Necrotic Ringspot**, OR.

Phomopsis leaf blight or blotch and Mycosphaerella (Phyllosticta) leaf spot are common and rather disfiguring on bushes in shade or under tree drip. In light cases, removal of spotted leaves is sufficient.

MOUNTAIN-MAHOGANY (*Cercocarpus*)
LEAF SPOT. **Septogloeum cercocarpi**, CA.
ROT, Wood. **Stereum hirsutum**, OR.

SPOT ANTHRACNOSE. **Sphaceloma cercocarpi**, CA.

MULBERRY (*Morus*)

BACTERIAL Scorch. **Xylem limiting bacteria**, Mid-Atlantic and Southern U.S; **Xylella fastidiosa**, CA.

BACTERIAL Hairy Root. **Agrobacterium rhizogenes**, NE.

BACTERIAL Leaf Spot. **Pseudomonas syringae** pv. **mori**, general.

BLIGHT, Berry; Popcorn Disease. **Ciboria carunculoides**, NC to FL, TX, VA.

BLIGHT, Twig. **Myxosporium diedickei**, TX, WA.

CANKER; Twig Blight. **Cytospora** sp., NJ, TX; **Dothiorella** sp.; **D. mori**, NJ, TX; **Gibberella baccata** var. **mori**, widespread; **Nectria** sp., widespread; **N. cinnabarina**, wide spread; **Sclerotinia** sp., TX.

LEAF SPOT. **Cercospora moricola**, PA to FL, NE, TX; **Cercosporella mori**, NE, OK, TX; **Exosporium** sp., FL; **Mycosphaerella arachnoidea**, false mildew, GA, NC; **M. mori**, widespread; **Phyllosticta moricola**, KS; **Cytospora** sp., VA.

NEMATODE, Root Knot. **Meloidogyne** spp., NJ, NC, OK.

POWDERY MILDEW. **Phyllactinia corylea**, OH; **Uncinula geniculata**, NY to KS, AL.

ROT, Heart. **Polyporus farlowii**, AZ, NM; **P. hispidus**, CT; **Ganoderma applanatum**; **Hymenochaete agglutinans**, MD.

ROT, Root. **Armillaria mellea**, DE, NC, OK; **Helicobasidium purpureum**, TX; **Phymatotrichum omnivorum**, TX.

ROT, Wood. **Schizophyllum commune**, CA; **Stereum cinerescens**, MA.

RUST. **Cerotelium fici** (II), LA.

Bacterial leaf spot damages nursery trees having overhead irrigation; Mycosphaerella leaf spot sometimes defoliates older trees.

MYRTLE (*Myrtus*)

LEAF SPOT. **Pestalotia decolorata**, LA.

ROT, Root. **Cylindrocladium pauciramosum**, CA.

ROT, Stem. **Sclerotium rolfsii**, FL.

RUST, **Puccinia vincae**, CA.

N

NANDINA

ANTHRACNOSE. **Glomerella cingulata**, TX.

LEAF SPOT. **Cercospora nandinae**, AL, SC.

NEMATODE, Root Knot. **Meloidogyne** sp., NC, TX.

NONPARASITIC. **Chlorosis**. Alkaline soil, TX.

POWDERY MILDEW. **Microsphaera berberidis** (**Erysiphe berberidis**), CA, Pacific Northwest.

ROT, Root. **Phymatotrichum omnivorum**, TX.

VIRUS. **Cucumber Mosaic**, MD, GA; **Nandina Mosaic**, CA; **Nandina Stem Pitting**, FL.

NECTARINE (*Prunus persica* var. *nectarina*)

BACTERIAL Canker. **Pseudomonas syringae**, CA.

BACTERIAL Crown Gall. **Agrobacterium tumefaciens**, MO.

BACTERIAL Leaf Spot; Canker. **Xanthomonas pruni**, OK.

BACTERIAL, MLO. **Peach X-Disease**.

CANKER. **Valsa leucostoma**, DC.

LEAF CURL. **Taphrina deformans**, CA, OR, TX, WA.

LEAF SPOT. **Cristulariella pyramidalis**, FL.

LEAF SPOT; Shot Hole. **Coryneum carpophilum**, CA, OR, WA.

NEMATODE, Lesion. **Pratylenchus thornei**, CA.

NEMATODE, Root Knot. **Meloidogyne** sp., CA; usually resistant.

POWDERY MILDEW. **Podosphaera oxyacanthae**, ID, WA; **Sphaerotheca pannosa**, ID, NY, WA.

ROT, Brown; Twig Blight. **Monilinia laxa**, CA, WA; **M. fructicola**, CT, NY, TX.

ROT, Pink Mold. **Trichothecium roseum**, CA.

ROT, Sour. **Geotrichum candidum**, CA; **Issatchenkia scutulata**, CA; **Kloeckera apiculata**, CA.

SCAB. **Cladosporium carpophilum**, CT, DE, IL, NY, PA, TX.

VIRUS. **Peach Mosaic**; **Peach Yellows**; **Peach Latent Mosaic Viroid**, WA; **Hop Stunt Viroid**, Ontario, Canada.

R.K. Horst, *Field Manual of Diseases on Trees and Shrubs*,

DOI 10.1007/978-94-007-5980-0_17,

NEMESIA

BACTERIAL, Proliferation. **Rhodococcus fascians**, OR.

VIRUS. Flower Break. **Angelonia angustifolia**, CA; **Hosta Virus X**, **Tobacco Rattle Virus**; **Helleborus Net Necrosis**; **Nemesia Ring Necrosis**, **Diascia Yellow Mottle**, Pacific Northwest.

NINEBARK (*Physocarpus*)

LEAF SPOT. **Cercospora spiraea**, IN; **Marssonina neilliae**, CA, TX, WI; **M. lonicerae**, OR; **Phyllosticta opulasteris**, ID; **Ramularia spiraeae**, MI, NY, WI.

POWDERY MILDEW. **Sphaerotheca macularis**, MA to WI.

ROT, Root. **Phymatotrichum omnivorum**, TX.

ROT, Wood. **Fomes conchatus**, NY.

O

OAK (Quercus)

ANTHRACNOSE; Leaf and Twig Blight. **Gnomonia quercina** (**Gloeosporium quercinum**), general.

BACTERIAL Canker. Unidentified, IL.

BACTERIAL Crown Gall. **Agrobacterium tumefaciens**, MD, MI.

BACTERIAL Drippy Nut. **Erwinia quercina**, CA, on live oak.

BACTERIAL Leaf Scorch. **Xylella fastidiosa**, DC, FL, GA, KY.

BACTERIAL Wetwood. **Erwinia nimipressuralis**, VA.

BLACK MILDEW. **Morenoella quercina**, SC to FL, GA, TX; **Irenina manca**, MS.

BLIGHT, Sudden Oak Death. **Phytophthora ramorum**, (also branch canker) CA, OK.

BLIGHT, Twig. **Cryptocline cinerescens**, CA; **Diplodia longispora**, NY to NC, IL, WI; **D. quercina**, CA; **Discula quercina**, CA.

CANKER, Bark. **Aleurodiscus oakesii**, NY to IL, CA, IA; **A. candidus**; **A. acerinus**; **A. griseo-canus**; **Dichaena quercina**, NJ.

CANKER, Basal. **Cryphonectria parasitica**, PA.

CANKER, Bleeding. **Phytophthora cactorum**, FL, MA, NY; **P. cinnamomi**, FL.

CANKER, Branch. **Endothia gyrosa**, VA.

CANKER, Branch; Decline; Dieback. **Botryodiplodia gallae**, MI; **Cephalosporium** sp.; **Hyalodendron** sp.

CANKER, Bot. **Diplodia corticola**, CA.

CANKER, Chestnut. **Endothia parasitica**, MS, SC, VA.

CANKER, Felt Fungus. **Septobasidium** spp., NC to FL, LA.

CANKER, Trunk. **Nectria galligena**; **Strumella coryneoides** (*Urnula craterium*); **Diplodia corticola**, CA.

CANKER, Twig. **Cytospora chrysosperma**, NJ, RI; **Endothia parasitica**, occasional; **Dothiorella quercina**, MD, VA; **Fusarium solani**; **Physalospora glandicola**, MD; **P. obtusa**, MN, VA; **P. rhodina**, VA; **Pseudovalsa longipes**; **Pyrenochaete venuta**; **Coryneum japonicum**, FL.

DECLINE. **Hypoxylon atropunctatum**, **H. punctulatum**, **H. mediterraneum**, VA; **Phytophthora quercina**, **P. cambivora**, MO.

R.K. Horst, *Field Manual of Diseases on Trees and Shrubs*,
DOI 10.1007/978-94-007-5980-0_18, © Springer Science+Business Media Dordrecht 2013

Dodder. **Cuscuta** spp., occasional in forest nurseries.

Leaf Blister. **Taphrina caerulescens**, Northeast to North central and Gulf states; also CA, CO, GA, UT, WY.

Leaf Spot. **Actinopelte dryina**, occasional; **Acantharia echinata**, black, on live oak; **Cercospora macrochaeta**, CA; **Cylindrosporium microspilum**, **C. kelloggii**, CA; **Ciborinia** (*Sclerotinia*) **candolleana**; **C. hirtella**; **Dothiorella phomiformis**, widespread; **Gloeosporium septorioides**; **Marssonina martini**, general; **Leptothyrella californicum**, on live oak, CA; **Monochaetia desmazierii**, widespread; **Mycosphaerella** sp.; **Phyllosticta** spp.; **Septogloeum defolians**, CA; **S. querceum**, WI; **Septoria** spp.; **Venturia orbicula**, NY to VA, OH.

Mistletoe. **Phoradendron serotinum (flavescens)**, NC to FL, TX; **P. villosum**, CA.; **P. tomentosum**, TX; **Psittacanthus macrantherus**, Sinaloa, Mexico.

Mold, Leaf. **Cladosporium brevipes**; **C. herbarum**.

Nematode, Dagger. **Xiphinema americanum**, Southeast.

Nematode, Ring. **Criconemoides annulatum**; **C. teres**; **Hemicriconemoides biformis**.

Nematode, Root. **Hoplolaimus coronatus**; **Pratylenchus** sp.; **Meloidogyne** sp.

Nonparasitic **Chlorosis**. Iron deficiency, especially in pin oaks.

Parasitic Lichen. **Strigula elegans** and **S. complanata**, Southern U.S., LA.

Powdery Mildew. **Erysiphe trina**, witches' broom, on live oak, CA; **Microsphaera alni**, widespread; **Phyllactinia corylea**, widespread; **Sphaerotheca lanestris**, "brown mildew," serious on coast live oak, AL, CA, MS, NC; **Saccardia quercina**, AZ, GA.

Rot, Heart. **Daedalea quercina**, widespread, other species; **Fistulina hepatica**; **Fomes** spp.; **Hericium erinaceous**; **Polyporus** spp.; **Stereum** spp.

Rot, Heart. **Phlebia chrysocrea**, eastern U.S. (PA to FL) and WI to MS.

Rot, Root. **Armillaria mellea**, widespread; **Clitocybe tabescens**, FL, MO, OK; **C. olearia**, CA; **Corticium galactinum** and other spp.; **Phytophthora cinnamomi**, of seedlings.

Rot, Root and Collar. **Phytophthora cinnamomi**, CA.

Rot, Wood. **Daldinia concentrica**; **D. vernicosa**; **Hypoxylon atropunctatum**; **Lentinus tigrinus**; **Linzites betulina** and other species; **Schizophyllum commune**; **Steccherinum ochraceum**.

Rust. **Cronartium quercuum** (*C. cerebrum*) II, III, widespread; 0, I on pines; **C. fusiforme** (II, III), southern states; **C. strobilinum** (II), AR, FL, IL, IA, KS, MS, MO; 0, I on long-leaf pines; **C. conigenum** (II, III), AZ; 0, I on pine cones.

Rust, Fusiform. **Cronartium fusiforme**, GA.

Spot Anthracnose. **Elsinoë quercus-falcatae**, on southern red oak, GA, NC; **E. quercicola**, FL.

Virus. **Oak Ringspot**, AR.

Wilt, Dieback. **Fusarium oxysporum** or **F. solani**, VA; also canker; **Phytophthora europa** (also yellowing and dieback), North Central U.S., MN, OH, PA, WV, WI.

Wilt, Oak. **Ceratocystis fagacearum** (*Chalara quercina*), AR, FL, IL, IN, IA, KS, KY, MD, MI, MN, MO, NE, NY, NC, OH, OK, PA, SC, TN, VA, WV, WI.

Witches' Broom. **Articularia quercina** var. **minor**. AZ, NM, UT.

Oak wilt is our most serious disease with red and black oaks often dying the first season symptoms appear. Anthracnose is general, most severe on white oak, defoliating in wet seasons. Leaf blister, important in the South, can be prevented by a single dormant spray. Powdery mildew, due to *Sphaerotheca lanestris*, is important in California, where it produces witches' brooms on live oaks. The honey mushroom, *Armillaria mellea*, sometimes called the oak fungus, causes "shoestring" root rot. Strumella canker is frequent in forest trees, sometimes found in ornamentals.

OCOTILLO, COACH-WHIP, CANDLEWOOD (*Fouquieria*)
ROT, Root. **Phymatotrichum omnivorum**, TX.
RUST. **Aecidium cannonii** (0, I), AZ.

OHIA (*Metrosideros* spp.); also *Eugenia*, *Syzgium* and *Psidium* spp. in Myrtaceae Family
RUST. **Puccinia psidii**, HI.

OLEANDER (*Nerium*)
ANTHRACNOSE, Leaf Spot. **Gloeosporium** sp., MA, MS, NJ, TX.
BACTERIAL Knot. **Pseudomonas syringae** pv. **tonelliana**, AZ, CA, CT.
BACTERIAL, Leaf Scorch. **Xylella fastidiosa**, FL, Costa Rica.
BLIGHT, Stem and Leaf. **Pseudomonas syringae**, CA.
CANKER; Witches' Broom. **Sphaeropsis** sp., FL.
DODDER. **Cuscuta indecora**, FL.
LEAF SCORCH. **Xylella fastidiosa**, LA.
LEAF SPOT. **Alternaria** sp. (*Macrosporium nerii*), AL, CA, FL, GA, MS; **Cercospora neriella**, also pod spot, AL, FL, LA, TX; **Phyllosticta nerii**, FL, LA, MI, TX; **Septoria oleandrina**, CA, FL, LA.
ROT, Root. **Clitocybe tabescens**, FL; **Phymatotrichum omnivorum**, TX.
ROT, Stem. **Calonectria crotalariae**, CA.
SOOTY MOLD. **Capnodium elongatum**, AL, FL.
SPOT ANTHRACNOSE, Scab. **Sphaceloma oleandri**, LA.

OLIVE (*Olea*)
ANTHRACNOSE. **Gloeosporium olivarum**, CA.
BACTERIAL Knot. **Pseudomonas syringae** pv. **savastanoi**, CA.
BLACK MILDEW; Leaf Spot. **Asternia oleina**, FL, GA.
DODDER. **Cuscuta indecora**, CA.
LEAF SPOT. **Cycloclonium oleaginum**, peacock spot, CA; Cercospora caldosporioides, also fruit spot, CA.
NEMATODE, Citrus. **Tylenchulus semipenetrans**.
NEMATODE, Lesion. **Pratylenchus musicola** (*P. coffeae*), CA.
NEMATODE, Root Knot. **Meloidogyne** sp., CA.
NONPARISITIC **Bitter Pit** and **Dry Rot** of fruit. Overnutrition
Exanthema. Dieback. Deficiency of organic matter.
Fruit Pit. Boron deficiency.
Soft Nose. Blue Nose. Moisture supply. On variety Sevillano only.
ROT, Root. **Armillaria mellea**, CA, TX; **Phymatotrichum omnivorum**, TX; **Polyporus olaea**, CA; **Fusarium solani**, Argentina.

Rot, Root, Twig Wilt. **Phytophthora nicotianae**, Argentina.
Virus. **Sickle Leaf**.
Wilt. **Verticillium albo-atrum**, CA.

ONCOBA
Rot, Root. **Phymatotrichum omnivorum**, TX.

ORANGE
See Citrus Fruits.

OSIER, BASKET-WILLOW (*Salix*)
Canker. **Cryptomyces maximus**, Blister; **Valsa salicina**, Twig, Branch, CA, IA.
Rust, Leaf. **Melampsora abieti-capraearum**, NY, PA.

OSOBERRY (*Osmaronia*)
Leaf Spot. **Cylindrosporium nuttallii**, CA to OR, WA; **Gloeosporium osmaroniae**, WA.
Powdery Mildew. **Phyllactinia corylea**, OR.

OXYDENDRON (Sourwood, Sorrel-Tree)
Blight, Twig. **Sphaerulina polyspora**, NC.
Leaf Spot. **Cercospora oxydendri**, AL, MS, TX, WV; **Mycosphaerella caroliniana**, GA, NC, TX, WV.
Rot, Root. **Phymatotrichum omnivorum**, TX.
Rot, Wood. **Poria punctata**, WV.

P

PACHISTIMA
LEAF SPOT. **Mycosphaerella pachystimae**, ID.

PACIFIC WAX MYRTLE (*Myrica californica*)
LEAF SPOT. **Phyllosticta myricae**, CA.
RUST. **Cronartium comptoniae** (II, III), OR.

PALM (*Chamaedorea*)
LEAF SPOT. **Bipolaris setariae**, FL; **Exserohilum** (*Helminthosporium*) **rostratum**, FL; **Phaeotrichoconis crotalariae**, FL.
ROT, Stem. **Gliocladium vermoeseni**, FL.

PALM ARECA (*Chrysalidocarpus*)
BLIGHT, Southern. **Sclerotium rolfsii**, CA.
LEAF SPOT. **Bipolaris setariae**, FL; **Exserohilum** (*Helminthosporium*) **rostratum**, FL; **Phaeotrichoconis crotalariae**, FL.

PALM, COCONUT (*Cocos*)
BACTERIAL, MLO. **Lethal Yellowing**, FL.
BLIGHT, Thread. **Pellicularia koleroga**, FL.
LEAF SCORCH; Leaf-Bitten Disease; Stem Bleeding. **Ceratocystis** (*Endoconidiophora, Thielaviopsis*) **paradoxa**, FL.
LEAF SPOT, Gray; Leaf Break. **Pestalotia palmarum**, secondary, FL.
ROT, Bud; Leaf Drop; Wilt. **Phytophthora palmivora**, FL; **Pythium** sp., wilt, FL.

PALM, DATE (*Phoenix*)
BLIGHT, Inflorescence; Fruit Rot. **Fusarium** spp., AZ.
CANKER. **Penicillium vermoeseni**, CA.
FRUIT SPOT, Brown. **Helminthosporium molle**, AZ; **Alternaria citri**, AZ, CA, TX.
LEAF SCORCH; Black Heart; Bud Rot. **Ceratocystis paradoxa**, AZ, CA.
LEAF SPOT. **Exosporium palmivorum**, Gulf states; **Pestalotia** sp., CA, FL; **P. palmarum**, FL; **Alternaria** sp., AZ, CA, FL, TX; **Annellophora phoenicis**, TX.

R.K. Horst, *Field Manual of Diseases on Trees and Shrubs*,
DOI 10.1007/978-94-007-5980-0_19, © Springer Science+Business Media Dordrecht 2013

LEAF SPOT; STEM LESIONS. **Colletotrichum fragariae**, FL.
NEMATODE, Root Knot. **Meloidogyne** spp., AZ, CA.
NONPARASITIC. **Black Nose**. Fruit checking from rain and high humidity. **Chlorosis**. Manganese deficiency, FL.
PHYTOPLASMA. **Witches' Broom**, GA, IL, IN, IA, OH.
ROT; Decline Disease. **Omphalia pigmentata** and **O. tralucida**, CA.
ROT, Fruit. **Aspergillus niger**, Calyx-end, CA, WA; **Catenularia fuliginea**, AZ, CA; **Penicillium roseum**, AZ, CA; **Pleospora herbarum**, CA; **Phomopsis phoenicola**, CA; **Alternaria stemphylioides**, CA.
ROT, Leaf-stalk; Shoot Blight; Fruit Rot. **Diplodia phoenicum**, AZ, CA.
ROT, Root. **Armillaria mellea**, CA; **Clitocybe tabescens**, FL; **Ceratostomella radiciola**, CA.
ROT, Wood. **Poria** spp., AZ, CA.
SMUT, False. **Graphiola phoenicis**, widespread.
WILT. **Fusarium oxysporum** f. sp. **canariensis**, LA, SC, TX; **F. oxysporum** f. sp. **palmarum**, FL.

PALM, FISHTAIL (*Caryota*)
BLIGHT, Leaf and Stem. **Glomerella cingulata**.
LEAF SPOT. **Bipolaris setariae**, FL; **Exserohilum** (*Helminthosporium*) **rostratum**, FL; **Phaeotrichoconis crotalariae**, FL.
PHYTOPLASMA. **Lethal Yellowing**, Puerto Rico.

PALM, FORSTER SENTRY (Sentry Palm) KENTIA (*Howeia*)
LEAF SPOT. **Calonectria colhounii**; **C. crotalariae**; **C. theae**, HI.

PALM, PINDO (*Butia*)
ROT, Root. **Armillaria tabescens**, SC.

PALM, QUEEN, PLUMY COCONUT (*Arecastrum*)
CANKER; Gummosis. **Dothiorella gregaria**, CA; **Penicillium vermoeseni**, CA.
LEAF SPOT. **Bipolaris setariae**, FL; **Exserohilum** (*Helminthosporium*) **rostratum**, FL; **Phaeotrichoconis crotalariae**, FL.
LEAF SPOT. **Exosporium palmivorum**, FL; **Glomerella cincta**, NJ; **Septoria cocoina**, MO.
LETHAL DECLINE, NECROSIS. **Fusarium oxysporum** f. sp. **palmarum**, FL.
NONPARASITIC. **Chlorosis**. Manganese deficiency, FL.
ROT, Bud; Wilt. **Phytophthora palmivora**, FL; **Pythium** sp., wilt.
ROT, Butt. **Ganoderma sulcatum**, FL.
SMUT, False. **Graphiola phoenicis**, FL.

PALM, RHAPIS (*Rhapis*)
LEAF SPOT. **Bipolaris setariae**, FL; **Exserohilum** (*Helminthosporium*) **rostratum**, FL; **Phaeotrichoconis crotalariae**, FL.

PALM, ROYAL (*Roystonea*)
ANTHRACNOSE; Petiole Spot. **Colletotrichum gloeosporioides**, FL, TX.
LEAF SPOT. **Alternaria** sp., FL; **Diplodia** sp., FL; **Epicoccum neglectum**, FL; **Helminthosporium** sp., leaf stripe, FL.

LITTLE LEAF. Cause unknown, FL.
NEMATODE, Root Knot. **Meloidogyne** sp., FL.
NEMATODE, Spiral. **Helicotylenchus nannus.**
PHYTOPLASMA. **Lethal Yellowing**, Puerto Rico.
SMUT, False. **Graphiola phoenicis**, FL.
WILT. **Phytophthora palmivora**, FL.

PALM, SUGAR (*Arenga*)
SMUT, False. **Graphiola phoenicis**, occasional.

PALM, WASHINGTON (*Washingtonia*)
BACTERIAL Leaf Spot. **Pseudomonas washingtoniae**, AZ.
LEAF SPOT. **Auerswaldia** sp., CA; **Cercospora** sp., FL; **Colletotrichum** sp., FL; **Cylindrocladium macrosporium**, FL; **Pestalotia palmarum**, FL; **Phoma palmicola**, secondary, TX.
LETHAL DECLINE; NECROSIS. **Fusarium oxysporum** f. sp. **palmarum**, FL.
NEMATODE, Root Knot. **Meloidogyne** spp., AZ, FL.
ROT, Bud. **Penicillium vermoeseni**, CA; **Phytophthora palmivora**, AZ, FL.
ROT, Root. **Clitocybe tabescens**, FL; **Phymatotrichum omnivorum**, FL.
ROT, Trunk. **Phytophthora** sp., CA.
SMUT, False. **Graphiola phoenicis**, FL, TX.
VIRUS. **Palm Mosaic**, CA. WILT. **Pythium** sp., FL.

The Penicillium bud rot causes serious losses to *Washingtonia filifera* in California; *W. robusta* is resistant.

PALMETTO, CABBAGE PALM (*Sabal*)
BLACK MILDEW. **Asterina sabalicola**, FL, GA; **Meliola** spp., Gulf states.
CANKER, Felt Fungus. **Septobasidium sabalis**, LA; **S. sabal-minor**, FL.
FRUIT DROP (Premature). **Colletotrichum pleosporioides**, FL.
LEAF SPOT. **Helminthosporium apiculiferum**, LA, MS; **Mycosphaerella serrulata**, FL, SC; **Phyllachora** (*Catacauma*) **sabal**, Black Spot, FL, GA, TX; **Phyllosticta palmetto**, LA, MS; **Myrianginella sabaleos**, Black Speck, FL, GA.
SMUT, False. **Graphiola congesta**, AL, FL, SC; **G. phoenicis**, FL, MS; **G. thaxteri**, FL.

PALOVERDE (*Cercidium*)
MISTLETOE. **Phoradendron californicum**, TX to CA.
ROT, Root. **Phymatotrichum omnivorum**, TX.

PANDANUS (Screw Pine)
LEAF SPOT. **Heterosporium iridis**, IA; **Macrophoma pandani**, CA, FL; **Melanconium pandani**, FL, MD; **Phomopsis** sp., NJ; **Pestalotia palmarum**, secondary, FL; **Volutella mellea**, NY.
NEMATODE, Burrowing. **Radopholus similis**, FL.

PAPAYA (*Carica*)
ANTHRACNOSE. **Colletotrichum capsici**, FL.

ANTHRACNOSE; Fruit, Stem, and Leaf Spot. **Glomerella cingulata**, general.
BLACK MILDEW. **Asterina caricarum**, FL.
BLIGHT, Flower; Leaf Spot. **Choanephora americana**, FL.
BLIGHT, Southern. **Sclerotium rolfsii**, TX.
DAMPING-OFF. **Pythium aphanidermatum**, CA, FL; **P. debaryanum**, TX; **Colletotrichum gloeosporioides**, HI; also seedling blight.
LEAF SPOT. **Asperisporium caricae**, FL; **Mycosphaerella caricae**, target spot, FL; **Phyllosticta caricae-papayae**, target spot, FL.
POWDERY MILDEW. **Golovinomyces cichoracearum**, CA; **Oidium caricae**, FL, TX.
NEMATODE. **Meloidogyne** sp., FL, TX.
ROT, Crown, Collar. **Calonectria** sp. (*Cylindrocladium*), HI.
ROT, Fruit. **Diplodia** sp., TX; **Colletotrichum magna**, Brazil.
ROT, Root. **Phymatotrichum omnivorum**, TX.
ROT, Stem. **Fusarium** sp., CA, TX.
VIRUS. **Tobacco Ring Spot**; **Papaya Mosaic**; **Papaya Ringspot**.

PARKINSONIA
LEAF SPOT. **Cylindrosporium parkinsoniae**, TX; **Phyllosticta parkinsoniae**, TX.
MISTLETOE. **Phoradendron californicum**, CA, TX.
ROT, Root. **Clitocybe tabescens**, FL; **Phymatotrichum omnivorum**, AZ, TX
SOOTY MOLD. **Capnodium** sp., TX.

PAULOWNIA (Princess-Tree)
GALLS, Root. **Meloidogyne arenaria, M. incognita, M. javanica, M. mayaguensis**, FL.
LEAF SPOT. **Ascochyta paulowniae**, MD; **Phyllosticta paulowniae**, AL, MD, NY, OK.
POWDERY MILDEW. **Phyllactinia corylea**; **Uncinula clintonii**.
ROT, Root. **Phymatotrichum omnivorum**, TX; **Armillaria mellea**.
ROT, Wood. **Polyporus spraguei**, AL; **P. versicolor**, MD; **P. robiniophilus**.

PAWPAW (*Asimina*)
BLOTCH, Leaf. **Phloeospora asiminae**, IL, KS, MO, OH, WV.
BLOTCH, SOOTY; FLYSPECK. **Colletogloeum** sp., **Dissoconium aciculare**, **Pelaster** sp. P21, **P. fructicola, Stomiopeltis** sp. RS1, **S. versicola**, and **Dothiodeomycete** sp. CS2, IA.
CANKER; DIEBACK. **Nectria cinnabarina**; **Polyporus amplectrens**, FL, GA; **Sphaeropsis asiminae**, MD, WV; **Valsa ambiens**, VA, WV.
LEAF SPOT. **Cercospora asiminae**, widespread; **Phyllosticta asiminae**, widespread; **Mycosphaerella** sp., IN, OH, WV; **Septoria asiminae**, TX.
PARASITIC LICHEN. **Strigula elegans** and **S. complanata**, Southern U.S., LA.
ROT, Root. **Phymatotrichum omnivorum**, TX.
ROT, Sapwood. **Poria isabellina**, VA, WV.

PEACH (*Prunus persica*)

ANTHRACNOSE. **Colletotrichum acutatum**, SC.

BACTERIAL Crown Gall. **Agrobacterium tumefaciens**, general.

BACTERIAL Hairy Root. **Agrobacterium rhizogenes**, general.

BACTERIAL Leaf Spot. **Xanthomonas pruni**, eastern, central, southern states.

BACTERIAL, MLO. **Peach X-Disease**, CT, MA, MI, NY, OH, PA.

BACTERIAL Shoot Blight; Canker; Gummosis. **Pseudomonas syringae**, CA, OR.

BLIGHT, Blossom; Green Fruit Rot. **Sclerotinia sclerotiorum**, CA.

BLIGHT, Leaf. **Fabraea maculata**, CA, NJ.

BLIGHT; Shot-Hole Disease; Pustular Spot. **Coryneum carpophilum**, general.

BLIGHT, Twig. **Coniothyrium** sp., TX; **Cyphella marginata**, OR; **Nectria cinnabarina**, AL.

CANKER, Crown. **Phytophthora citrophthora**, CA; **P. cactorum**, stem, AR, CA; **P. syringae**, OR.

CANKER; DIEBACK; Gummosis. **Botryosphaeria ribis** var. **chromogena**, FL, GA; **B. dothidea**, AL, FL, GA, LA, TN, TX; **Valsa** (*Cytospora*) **cincta** and **V. leucostoma**, widespread East and central states; **Coniothrium fuckelii**, WV; **Ceratocystis fimbriata**, CA.

CANKER; Peach; Constriction Disease. **Fusicoccum amygdali**, East and South. Sometimes reported as **Phoma persicae**.

DAMPING-OFF. **Rhizoctonia solani**, AR, CT.

DIEBACK. **Phomopsis** sp., AL.

LEAF CURL. **Taphrina deformans**, general.

LEAF SCORCH, Stunting, Branch Dieback. **Xylella fastidiosa**, NM.

LEAF SPOT. **Phyllosticta circumscissa**, widespread; **P. persicae**, MD, NE, OH; **Mycosphaerella persicae**, frosty mildew, widespread.

LEAF SPOT; Shot-Hole. **Cercospora circumscissa**, general; **C. consobrina**, IL, LA.

NEMATODE. **Tylenchusus palustris**, AL, AR, GA, VA.

NEMATODE, Dagger. **Xiphinema americanum**; **X. pacificum**, GA.

NEMATODE, Lesion. **Pratylenchus penetrans**; **P. vulnus**.

NEMATODE, Ring. **Criconemoides similis**, **C. xenoplax**, NJ.

NEMATODE, Root Knot. **Meloidogyne incognita**; **M. javanica**, on "nematode-resistant" varieties; **M. arenaria**.

NEMATODE, Sting. **Belonolaimus gracilis**.

NONPARASITIC. **Catface**. Insect blemish on fruit, often from plant bugs.

Chlorosis. Iron and magnesium deficiency in alkaline soils, Southwest.

Gummosis. Winter injury, bad drainage.

Internal Bark Necrosis. Manganese toxicity, IL.

Little Leaf. Zinc deficiency, CA.

Stem Pitting. Cause unknown, WV.

Suture Spot. Cause unknown. Decline, PA.

PHYTOPLASMA. Peach Rosette. Ontario Canada.

POWDERY MILDEW. **Podosphaera oxyacanthae**, general; **Sphaerotheca pannosa** var. **persicae**, general.

Rot, Brown. **Monilinia fructicola**, general, SC.

Rot, Brown; Blossom and Twig Blight. **Monilinia laxa**, Pacific states.

Rot, Bud and Twig Blight. **Fusarium lateritium**, other spp., CA, GA, KS.

Rot, Fruit. **Botryosphaeria obtusa**, AL, GA; **B. dothidea**, AL, GA, SC; **Aspergillus niger**, black mold, cosmopolitan; **Botrytis cinerea**, gray mold, cosmopolitan; **Cephalothecium roseum**, pink mold, widespread; **Choanephora persicaria**, in market; **Diplodia natalensis**, gumming disease; **D. persicae**, GA; **Fusarium** spp., CA, IL, NY, TX; **Gibberella persicaria**, CA; **Gilbertella persicaria**, SC; **Glomerella cingulata**, occasional, AL, GA; **Rhizopus nigricans**, cosmopolitan; **Phomopsis** sp., NJ.

Rot, Pink Mold. **Trichothecium roseum**, CA.

Rot, Root. **Armillaria mellea**, cosmopolitan; **Clitocybe monadelpha**, AR, FL, MO, OK; **C. parasitica**, OK; **C. tabescens**, FL, GA; **Ganoderma curtisii**, NC, VA; **Phymatotrichum omnivorum**, AZ, TX; **Cylindrocladium floridanum**, GA.

Rot, Root and Crown. **Phytophthora megasperma**, NY, OH; **P. cryptogea**, NY, OH; **P. cactorum**, NY, OH.

Rot, Seedling Stem. **Sclerotium rolfsii**, South; wilt, **Cylindrocladiu floridanum**; **Penicillium** sp.

Rot, Silver Leaf. **Stereum purpureum**, occasional.

Rot, Sour. **Geotrichum candidum**, CA, SC; **Issatchenkia scutulata**, CA; **Kloeckera apiculata**, CA.

Rot, Wood. **Fomes** spp.; **Lenzites saepiaria**; **Polyporus hirsutus**, **P. lacteus** and **P. versicolor**, cosmopolitan; **Stereum hirsutum**; **Shizophyllum commune**, wound rot.

Rust. **Tranzschelia discolor** (II, III), general; 0, I on anemone, hepatica.

Scab. **Cladosporium carpophilum**, general.

Sooty Mold; Fruit Stain. **Fumago vagans**, cosmopolitan.

Virus. **Peach Asteroid Spot,** OK and TX to southern CA, OR, UT, WA; **Peach Calico**, ID, WA; **Peach Dwarf**; **Peach Golden Net**, CO; **Peach Little Peach**, eastern U. S., CT to NC to MI, MO; **Peach Mosaic**, OK, TX to CA, CO, UT; **Peach Mottle**, ID; **Peach Necrotic Leaf Spot,** MI, Puebla, Mexico; **Peach Phony Disease**, Southeast to MO, TX; **Peach Red Suture**, IN, MD, MI; **Peach Ring Spot,** CA, WA, **Peach Rosette**, SC to FL, MS, TN, MI to AR, OK; **Peach Rosette Mosaic**, MI, NY; **Peach Stubby Twig**; **Peach Wart**, AZ, CA, ID, OR, WA; **Peach Yellow Bud Mosaic**, CA, **Peach Yellow Leaf Roll** (Western X-Disease, Cherry Buckskin), CA, CO, ID, OR, UT, WA; **Peach Yellows; Mule's Ear Disease** (Drake Almond Bud Failure); **Prunus Ring Spot**; **Cherry Yellows**; **Green Ring Mottle; Prunus Necrotic Ring Spot,** CA; **Tomato Ring Spot,** CA; **Peach Latent Mosaic Viroid**, WA; **Hop Stunt Viroid**, Ontario, Canada; **Plum Pox**, PA; **Stem Pitting; Fravirus carolinianan, Baccharis halimifolia,** VA.

Wilt. **Verticillium albo-atrum**, occasional.

Brown rot is the No. 1 peach enemy, and spray schedules are built around it, although they start with a dormant spray for leaf curl. Get the latest advice and spray schedule from your county agricultural agent. Peach foliage is sensitive to arsenicals; do not allow a spray prepared for shade trees to touch peaches. The commercial propagator must take many precautions to provide stock free from the numerous viruses.

PEAR (*Pyrus*)

ANTHRACNOSE, Northwestern. **Neofabraea malicorticis**, OR, WA.

BACTERIAL Blossom, Twig Blight; Canker. **Pseudomonas syringae**, AR, CA, CT.

BACTERIAL Crown Gall. **Agrobacterium tumefaciens**, general.

BACTERIAL Fire Blight. **Erwinia amylovora**, general.

BACTERIAL Fruit Rot. **Erwinia carotovora**, MA.

BLIGHT, Leaf; Black Fruit Spot. **Fabraea maculata**, general.

BLIGHT, Thread. **Pellicularia koleroga**, NC to FL, TX, WV.

BLIGHT, Twig. **Corticium salmonicolor**, FL, LA; **Fusarium** spp., occasional; **Phomopsis ambigua**, widespread; **Valsa leucostoma**, WA.

BLOTCH, Sooty. **Gloeodes pomigena**, eastern states to OK, TX.

CANKER. **Botryosphaeria dothidea**, AL; **Cytospora** spp., OR, VA, WA; **Nectria cinnabarina**, coral spot; dieback; **N. galligena**, trunk canker; **Nummularia discreta**, DE, IA.

CANKER, Bark. **Glutinium microsporum**, OR; **Helminthosporium papulosum**, blister canker, black pox; **Myxosporium corticola**, NY to MI, MS, OR.

CANKER, Felt Fungus. **Septobasidium** spp., NC to FL, TX.

CANKER, Perennial. **Neofabraea perennans**, OR, WA.

DECLINE, Pear. **Candidatus Phytoplasma pyri**, Ontario, Canada.

FRUIT SPOT; Flyspeck. **Leptothyrium pomi**, eastern states.

LEAF SPOT. **Cercospora minima**, FL to TX; **C. pyri**, MI; **Coniothyrium pyrinum**, MA to AL, IA, TX; **Coryneum foliicola**, IN; **Hendersonia cydoniae**, NY; **Phyllosticta pyrorum**, IL, MS, SC; **Mycosphaerella sentina**, ashy leaf spot, fruit spot, widespread, especially in East.

MISTLETOE. **Phoradendron serotinum (flavescens)**, AZ, TX.

MISTLETOE, European. **Viscum album**, CA.

NEMATODE, Lesion. **Pratylenchus minyus**; **P. pratensis**, CA.

NEMATODE, Root Knot. **Meloidogyne** sp., CA.

NONPARASITIC. **Bitter Pit**. Moisture irregularity, Pacific Coast, NY.

Black End; **Hard End**. Oriental pear rootstocks on shallow, poor soil.

Black Leaf; **Brown Bark Spot**; **Brown Blotch**, of Kieffer fruit. Undetermined.

Chlorosis. Mineral deficiency, soil alkalinity, Pacific Coast.

Cork; Drought Spot; Fruit Pitting. Boron deficiency, Pacific Coast, TX.

Exanthema. Copper deficiency, CA, FL.

Little Leaf; Rosette. Zinc deficiency, sometimes boron, CA.

Marginal Leaf Blight; Scorch. Potassium or calcium deficiency, ID, WA.

Red Leaf. In Oriental pear, undetermined cause.

Scald. Immaturity; deficient ventilation.

Stigmonose. Insect punctures during growth, widespread.

Target Canker, Measles. Undetermined, CA, GA, NY, VA, WA.

POWDERY MILDEW. **Podosphaera leucotricha**, CO, MD, OR, WA; **P. oxyacanthae**, NJ.

Postharvest Disease. **Sphaeropsis pyriputrescens**, MI.

Rot, Bitter; Twig, Branch Canker. **Glomerella cingulata**, widespread but not destructive.

Rot, Black; Canker; Leaf Spot. **Physalospora obtusa**, widespread.

Rot, Brown. **Monilinia fructicola**, eastern states, TX, WA.

Rot, Brown; Blossom Blight. **Monilinia laxa**. Pacific states.

Rot; Blossom and Twig Blight. **Botrytis cinerea**, widespread.

Rot, Collar, Root. **Phytophthora cactorum**, widespread; **P. citrophthora**, CA.

Rot, Fruit. **Alternaria** sp., black mold; **Aspergillus** sp., cosmopolitan; **Botryosphaeria ribis**, black rot, VA; **Cephalosporium carpogenum**, storage; **Trichothecium roseum**, occasional pink mold; **Cladosporium** sp., occasional; **Coprinus urticicola**, OR; **Gloeosporium** sp., widespread **Neurospora sitophila**, ripe rot, NC; **Penicillium** spp., blue mold, widespread; **Phialophora malorum**, storage; **Phoma exigua**; **P. mali**; **Rhizopus nigricans**, black mold, cosmopolitan; **Pleospora fructicola**; **Sclerotinia sclerotiorum**, CA, WA; **Sporotrichum malorum**, storage rot, OR, WA; **Phacidiopycnis washingtonensis** (also Canker and Blight), WA; **Sphaeropsis pyriputrescens**, WA; **Potebniamyces pyri**, WA.

Rot, Heart. **Fomes igniarius**; **F. pinicola**; **Polyporus** spp.

Rot, Root. **Armillaria mellea**, widespread; **Phymatotrichum omnivorum**, AZ, OK, TX; **Xylaria** sp., ID, IN; **Xylaria mali**, VA; **Phytophthora cinnamomi**.

Rot, Silver Leaf. **Stereum purpureum**, NY, OR; **S. hirsutum**, trunk rot.

Rot, Stem, Calyx. **Phacidiopycnis piri** (also canker and postharvst fruit rot), OR.

Rot, Trunk. **Schizophyllum commune**, cosmopolitan.

Rust. **Gymnosporangium clavariiforme** (0, I), leaves and fruit, SC: **G. clavipes** (0, I), chiefly on fruit; **G. fuseum**, CA; **G. globosum** (0, I), on leaves and fruit, eastern states to IA, MN; **G. hyalinum** (0, I), FL; **G. kernianum** (0, I), on leaves and fruit, AZ, CO; **G. nootkatense** (0, I), on Asiatic pear, AK; **G. sabinae**, MI.

Scab. **Venturia pyrina**, general.

Spot Anthracnose, Scab. **Elsinoë pini**, OR, WA.

Viroid. **Pear Blister**. Argentina.

Virus. **Apple Chlorotic Leaf Spot, Spy Epinasty Decline**, **Virginia Crab Stem Pitting and Brownline**, **Apple Mosaic**, **Flat Limb**, **Pear Vein Yellows/Red Mottle**, **Pear Ring-Pattern Mosaic**, **Pear Latent** on Bradford Pear (*Pyrus calleryana*), MD; **Pear Stony Pit**, Pacific Coast states, NY; **Pear Decline**, CA, OR, WA; **Pear Leaf Curl**, CA; **Pear Bark Measles**, CO; **Apple Stem Pitting**, Argentina.

Fire blight is the limiting factor in pear production, many orchards having been abandoned because of this devastating bacterial disease. Kieffer pears and some Asiatic varieties are resistant and are being used in breeding; some resistant forms are now available. Sprays during blossoming, usually an antibiotic, sometimes combined with weak copper, are used in conjunction with sanitary measures for fireblight control. Pear decline, a virus disease transmitted by the pear psylla, is serious on the Pacific Coast.

PEA-TREE (*Caragana*)
BACTERIAL Hairy Root. **Agrobacterium rhizogenes**, KY to NE, OK.
BLIGHT, Leaf. **Ascochyta** sp., OH; **Septoria** sp., MN.
BLIGHT, Pod. **Botrytis cinerea**, MA.
DAMPING-OFF. **Rhizoctonia solani**, ND.
LEAF SPOT. **Phyllosticta gallarum**, AK, WI.
POWDERY MILDEW. **Microspora (Erysiphe) palczewskii**, AK, ID, MN, ND, WA; **Erysiphe arborescens,** WI, Quebec, Canada.
ROT, Root. **Pellicularia filamentosa**, ND; **Phymatotrichum omnivorum**, TX; **Phytophthora cactorum**, wilt, of seedlings.

PECAN (*Carya illinoensis*)
ANTHRACNOSE. **Glomerella cingulata**, GA.
BACTERIAL Crown Gall. **Agrobacterium tumefaciens**, widespread.
BLIGHT, Thread. **Pellicularia koleroga**, FL, NC.
BLOTCH, Leaf. **Mycosphaerella dendroides**, South.
CANKER. **Cytospora** sp., AZ.
CANKER; Black Bark Spot. **Myriangium tuberculans**, GA, MS.
CANKER; DIEBACK. **Botryosphaeria berengeriana**, AZ, SC, to FL.
CANKER, Stem. **Microcera** (*Fusarium*) **coccophila**, LA, TX.
LEAF SCORCH, FUNGAL. **Xylella fastidiosa**, **Pestalotia**, **Epicoccum**, **Curvularia**, **Fusarium**, LA.
LEAF SPOT. **Microstroma juglandis**, GA to TX; **Phyllosticta convexula**, OK; **Septoria caryae**, TX; **Pestalotia uvicola**, FL, TX.
LEAF SPOT, Algal. **Cephaleuros virescens**, FL.
LEAF SPOT, Brown. **Cercospora fusca**, prevalent through pecan belt.
LEAF SPOT; Downy Spot. **Mycosphaerella caryigena**, GA and FL to TX.
LEAF SPOT; Liver Spot. **Gnomonia caryae** var. **pecanae**, AL, AR, GA, LA, MS, TX.
LEAF SPOT; Vein Spot. **Gnomonia nerviseda**, AR, IL, LA, MS, TX.
LEAF SPOT; Zonate. **Cristulariella pyramidalis**, AL.
MISTLETOE. **Phoradendron serotinum** (**flavescens**), widespread Gulf states.
NEMATODE. **Paratylenchus** sp.; **Trichodorus** sp.
NEMATODE, Awl. **Dolichodorus obtusus**.
NEMATODE, Dagger. **Xiphinema americanum**, Southeast.
NEMATODE, Root Knot. **Meloidogyne** spp.; **M. incognita**; **M. partityla**, FL, GA.
NONPARASITIC. **Black Pit**; **Kernel Spot**. Feeding punctures of plant bugs.
Leaf Scorch. Low fertility and soil moisture capacity.
Rosette. Zinc deficiency.
Sand Burn of Seedlings. High temperature.
POWDERY MILDEW. **Microsphaera alni**, occasional.
ROT, Heart. **Schizophyllum commune**, after drought injury, OK.
ROT, Kernel and Shuck. **Phytophthora cactorum**, GA.
ROT, on nuts. **Aspergillus chevalieri**, storage mold; **Trichothecium roseum**, pink mold.

Rot, Root. **Armillaria mellea**, CA; **Helicobasidium purpureum**, TX; **Phymatotrichum omnivorum**, TX; **Pestalotia uvicola**, FL, TX; **Clitocybe tabescens**, GA.
Scab. **Cladosporium effusum**, general.
Spot Anthracnose; Nursery Blight. **Elsinoë randii**, southeastern and Gulf states.
Virus. **Bunch Disease**. Undetermined.

Pecan scab is the limiting factor in nut production in southeastern states, with strains of the fungus sometimes attacking varieties long considered immune; four or five sprays are required for control. In Arizona severe infection from Phymatotrichum root rot has come where trees are intercropped with lucerne, which provides a rapid transit medium for the fungus. Getting rid of the lucerne and treating with ammonium sulfate saves some trees. Zinc sulfate, added to the soil or to foliar sprays, controls rosette.

PEPPER VINE, CISSUS (*Ampelopsis arborea*)
Blight, Thread. **Pellicularia koleroga**, LA.
Dodder. **Cuscuta compacta**, FL.
Leaf Spot. **Cercospora arboreae**, TX; **C. vitis**, LA; **Guignardia bidwellii** f. sp. **parthenocissi**, MS, NJ.

PERIWINKLE, MADAGASCAR (*Catharanthus*)
Blight, Foliar. **Phytophthora tropicalis**, VA.
Blight, Web. **Rhizoctonia solani**, LA; **Sclerotium rolfsii**, LA.
Phytoplasma, Yellowing and Witches' Broom. **Candidatus brasiliense**, Brazil.

PERSIMMON (*Diospyros*)
Anthracnose. **Colletotrichum acutatum**, NC.
Anthracnose; Fruit, Twig, Blight. **Gloeosporium diospyri**, East and South to KS.
Bacterial Crown Gall. **Agrobacterium tumefaciens**, CA.
Bacterial, Witches' Broom. **MLO**, TX.
Blight, Thread. **Pellicularia koleroga**, FL, on Japanese persimmon.
Blight, Twig. **Phoma diospyri**, FL, SC; **Physalospora obtusa**, Gulf states.
Blotch, Leaf. **Mycosphaerella diospyri**, Gulf states.
Canker; Dieback. **Diplodia natalensis**, AL, TX; **Botryosphaeria ribis**, AL, LA, MD.
Fruit Spot; Flyspeck. **Leptothyrium pomi**, FL; **Macrophoma diospyri**, AL, NJ, TX; **Dissoconium mali**, IA, China.
Fruit Spot, Algal. **Cephaleuros virescens**, FL.
Fruit Spot, Brown. **Cercospora diospyri**, IL, ME, MS, SC, TX, VA; Black, **C. fuliginosa**, AL, FL, GA, IL, MS, TX; **Pestalotia** sp., secondary; **Phyllosticta** sp., FL, IN; **Ramularia** sp., FL; **Fusicladium levieri**, CT, FL, MS.
Mistletoe. **Phoradendron serotinum (flavescens)**, FL, TX; **Viscum album**, CA, on *Diospyros kakai* (Japanese Persimmon).
Nematode, Citrus. **Tylenchulus semipenetrans**, CA.
Nematode, Root Knot. **Meloidogyne** sp., FL, SC, TX.
Powdery Mildew. **Podosphaera oxyacanthae**, TX.
Rot, Fruit. **Alternaria** sp., occasional on Japanese persimmon; **Botrytis cinerea**, gray mold; **Penicillium expansum**, blue mold, cosmopolitan; **Physalospora** spp., AL, AZ, GA, NY, TX; **Rhizopus stolonifer**, TX.

Rot, Root. **Armillaria mellea**, CA; **Phymatotrichum omnivorum**, TX.
Rot, Wood. **Daedalea ambigua**, LA, MS; **Daldinia concentrica**, LA, VA; **Fomes** spp.; **Hericium erinaceus**, LA; **Lentinus tigrinus**, LA, MS; **Pleurotus ostreatus**, LA, MS; **Polyporus** spp.; **Schizophyllum commune**, KY, TN.
Wilt. **Cephalosporium diospyri**, AL, AR, FL, GA, MS, NC, OK, SC, TN, TX; **Verticillium albo-atrum**, TX.

Persimmon wilt started in Tennessee in 1933, and at the end of 5 years only 5 % of the persimmons in that native stand were alive. Oriental persimmons are resistant.

PHOENIX-TREE (*Firmiana simplex*)
Blight, Web. **Rhizoctonia microsclerotia**, FL.
Canker; Coral Spot. **Nectria cinnabarina**, OK.
Rot, Root. **Phymatotrichum omnivorum**, TX.

PHOTINIA (Christmasberry, Toyon and Oriental Species)
Anthracnose. **Gloeosporium** sp., MS.
Bacterial Fire Blight. **Erwinia amylovora**, CA, NJ.
Blight, Leaf. **Fabraea maculata**, CA, LA.
Dieback. **Cytospora** sp.
Leaf Spot. **Cercospora** sp., GA; **C. heteromeles**, CA, TX; **C. photinia-serrulata**; **C. eriobotryae**; **Lophodermium heteromeles**, CA; **Phyllosticta heteromeles**, CA, TX; **Pestalotia** sp., NJ; **Septoria photiniae**, CA, PA.
Powdery Mildew. **Podosphaera leucotricha**, CA; **Sphaerotheca pannosa**, OR.
Rot, Crown. **Phytophthora cactorum**, CA.
Rot, Root. **Clitocybe tabescens**, FL; **Phymatotrichum omnivorum**, TX; **Phytophthora lateralis**, NC; **P. syringae**, NC.
Rust. **Gymnosporangium japonicum** (0, I), III on juniper; **G. clavipes**, MA.
Scab. **Photinia photinicola** (*Spilocaea eriobotryae*), on leaves, berries, CA.

PIERIS (Mountain, Japanese Andromeda)
Blight, **Phytophthora ramorum**, OR; **P. tropicalis**, VA.
Blight, Twig. **Cytospora** sp., OR.
Dieback. **Phytophthora** sp., PA; **Phytophthora cambivora**, NC.
Leaf Spot. **Pestalotia** sp., CT, NJ; **Phyllosticta andromedae**, NJ; **P. maxima**, CT, NJ; **Alternaria tenuis**, RI.
Leaf Spot, Tar. **Rhytisma andromedae**, FL, GA.
Rot, Root. **Armillaria mellea**, NJ.

PIGEON PEA (*Cajanus*)
Phytoplasma, **Witches' Broom**. FL.
Virus. **Pigeon Pea Golden Mosaic**.

PINE (*Pinus*)
Black Mildew. **Lembosia acicola**, CA.
Blight, Diplodia. **Diplodia pinea**, AL, LA, MS.
Blight, Brown Felt. **Herpotrichia nigra**, Northwest, on snow-buried foliage at high altitudes; **Neopeckia coulteri**, CA, MT.

BLIGHT, Brown Spot Needle. **Systremma** (*Scirrhia*) **acicola**, on hard pines in South, on Red pine, WI.

BLIGHT, Foliage. **Rhizoctonia solani**, VA.

BLIGHT, Needle. **Cytospora pinastri**, ME, NJ, PA; **Hendersonula pinicola**, NC, TN, WY; **Pullularia pullulans**, after insect injury; **Dothistroma pini**, IL, IA, NE, OH, OK, MD; **Septoria spadicea**, MN, NH, NY, VT, VA; **Lecanosticta** sp.; **Dothistroma pini**, MN, MT, PA; **D. septospora**, PA, VT; **Lophodermella cerina**, AL, FL, LA, MS.

BLIGHT, Seedling. **Botrytis cinerea**, cosmopolitan; **Cylindrocladium scoparium**, NJ; **Fusarium subglutinans**, AL, NC, seedling mortality; **Rhizina undulata**, CA, ME, MN, northern Rocky Mts.; **Thelephora terrestris**, KS, ME, NJ, OH, PA, northern Rocky Mts.; **Scleroderris lagerbergii**, MI, NY, WI, also canker.

BLIGHT, Shoot. **Fusarium moniliforme** var. **subglutinans**, FL, GA; **Diplodia gossypina**, GA; **Gremmeniella abietina**, NH, NY; **Sphaeropsis sapinea**, NE; **Diplodia pinea**, TX, WI.

BLIGHT, Snow. **Phacidium infestans**, occasional, New England; **P. convexum**, NC.

BLIGHT, Tip; Twig; Collar Rot. **Diplodia pinea** (*Sphaeropsis ellisii* and *S. Sapinea*), New England to CA, IA, KS, ND, SD, VA, WI. **Cenangium ferruginosum** (*C. abietis*), "pruning twig blight," widespread, sometimes secondary; **Monochaetia pinicola**.

CANKER. **Atropellis pinicola**, Northwest; **A. piniphila**, Northwest, AL, NM, TN; **A. tingens**, GA, MA, NH, NC, PA, VA; **A. arizonica**, AZ; **Caliciopsis pinea**, New England to SC, TN; **Dasyscyphus ellisiana**, on introduced pines in eastern states; **D. pini**, on five-needle pines, Pacific Northwest, MI; **Nectria** spp., New England to NC, IA; **Rhabdospora mirabilissima**, of seedlings, NY; **Tympanis confusa**, VA.

CANKER. **Scleroderris lagerbergii** on red pine MI, MN, NY, VT, WI.

CANKER, Bark. **Aleurodiscus amorphus**, widespread; Felt, **Septobasidium** spp.

CANKER, Bleeding. **Sphaeropsis ellisii**, PA.

CANKER, Pitch. **Fusarium lateritium** f. sp. **pini**, FL, VA; **F. moniliforme** var. **subglutinans**, FL, NC; **F. circinatum**, AL, CA.

DAMPING-OFF. **Rhizoctonia solani** and **Pythium** spp., cosmopolitan; **Fusarium moniliforme** var. **intermedium**, GA.

DECLINE. **Helicotylenchus multicinctus**, GA.

DIEBACK, Branch. **Diplodia pinea**, WI.

DIEBACK, Tip. **Sirococcus strobilinus**, CA, MI, MN, WI.

MISTLETOE. **Psittacanthus macrantherus**. Sinaloa, Mexico; **Arceuthobium blumeri** and **A. globosum** subsp. **globosum**, Mexico.

MISTLETOE, Dwarf. **Arceuthobium americanum**, Rocky Mt. states; Pacific Northwest; **A. campylopodum**, western dwarf, Rocky Mt. states to Pacific Coast, TX; **A. vaginatum**, southern Rocky Mt. states, Mexico; **A. vaginatum** subsp. **cryptopodum**, CO; **A. cyanocarpum**, CA, CO, OR, WY; **A. occidentale**, CA; **A. gilli**, AR; **A. vaginatum** subsp. **vaginatum**, AR; **A. hondurense**, Mexico; **Dwarf Mistletoe, A. aureum** subsp. **aureum,** Mexico; **Cladocolea cupulata** on **P. duglasiana** and **P. herrarie**, Mexico (Durango and Sinaloa).

Needle Cast. **Bifusella linearis**, tar spot, widespread; **B. striiformis**, CA; **Elytroderma deformans**, also Witches' Broom, SD to AZ, CA, GA, WA; **Hypoderma desmazierii**; ME to NC, GA, WI; **H. hedgecockii**, Southeast; **H. lethale**, gray blight, on hard pines, New England to FL, LA; **H. pedatum**, CA; **H. pini**, CA, NV; **H. saccatum**, tar spot, CO, NM; **Hypodermella arcuata**, CA, OR; **H. ampla**, tar spot, Great Lakes states; **H. cerina**, CA; **H. concolor**, CO to ID, MT, OR; **H. lacrimformis**, CA, OR; **H. limitata**, CA; **H. medusa**, CA, CO; **H. montana**, CA to ID, OR; **H. montivaga**, CA to MT, OR; **Lophodermium nitens**, New England to GA and MI and Pacific Northwest, on five-needle pines; **L. pinastri**, widespread; **L. ponderosae**, CA; **Canavirgella banfieldii**, MI.

Needle Cast. **Bifusella saccata**, CA, CO; **Hypodermella** sp., CA, OR; **Lophodermium durilabrum**, OR; **L. seditiosum**, MI.

Needle Droop. Abiotic inability of poor roots to acquire sufficient water, MI, MN, WI.

Nematode, Dagger. **Xiphinema americanum**.

Nematode, Lance. **Hoplolaimus coronatus**.

Nematode, Leaf. **Aphelenchoides fragariae**, FL.

Nematode, Pinewood. **Bursaphelenchus xylophilus**, vectored by pine sawyer beetle, IL, MO; also **B. lignicolus**, MO.

Nematode, Ring. **Criconemoides lobatum**; FL; **C. similis**, FL, NC.

Nematode, Root Knot. **Meloidogyne** sp., NM; **Meloidodera javanica**, FL; **M. floridensis**, on slash pine, FL.

Nematode, Sting. **Belonolaimus gracilis**.

Nonparasitic. **Ozone Injury**; **Sulfur Dioxide Injury**; IN, NJ, WI; **Needle Curl**. High temperature; **Chlorotic Dwarf**.

Rot, Charcoal. **Macrophomina phaseoli**.

Rot, Heart; Wood. **Fomes** spp., general; **Fomes annosus**; **Lentinus lepideus**, widespread; **Polyporus** spp., widespread; **Poria** spp., **Lenzites saepiaria**; **Stereum** spp.; **Trametes** spp., widespread.

Rot, Hypocotyl. **Fusarium oxysporum**, CA; **F. oxysporum** var. **redolens**, MN, root lesions; **Longidorus** sp. (root damage and seedling stunting), GA.

Rot, Root. **Armillaria mellea**, widespread; **A. gallica**, HI; **Ceratocystis wageneri**, OR, Pacific Northwest; **Clitocybe tabescens**, FL; **Cylindrocladium scoparium**, seedling blight, NJ, PA, WA; **C. clavatum**, FL; **Fusarium solani**, PA; **F. oxysporum**, PA; **F. avenaceum**, PA; **Phytophthora cactorum**, seedling blight, Northeast; **Phymatotrichum omnivorum**, TX; **Sparassis radicata**, MT to OR, WA; **Verticicladiella wagenerii**, CA, MT, western states; **V. procera**, FL, MN; **V. penicillata**, ID; **V. abietina**, ID.

Rot, Root. **Fomes annosus**, VA, WI; **Polyporus schweinitzii**, VA; **Phytophthora parasitica** var. **nicotianae**, CA; **P. citricola**, CA; **P. dreschleri**, NC; **Iononutus circinatus**, FL.

Rot, Root; Little Leaf. **Phytophthora cinnamomi**, NC, TN, VA, Dominican Republic.

Rot, Root; blue-stain fungi associated with roots; **Leptographium terebrantis**; **L. procerum**; **Ophiostoma** sp., **Graphium** spp., AL, TX.

Rot, White. **Amylostereum areolatum**, NY, eastern Canada.

Rust. **Coleosporium crowellii**, CA, NV, UT, NV; **Melampsora laricipopulina** (0, I), WA; II, III, on poplar; **M. medusae** f. sp. **deltoidae** (0, I), WA; II, III, on poplar.

Rust, Blister. **Cronartium ribicola, AZ.**

Rust, Comandra Blister. **Cronartium comandrae** (0, I), swellings in twigs, trunks of hard pines; II, III on bastard toadflax, ND.

Rust, Cone. **Cronartium conigenum**, hypertrophy of cones, especially in South; II, III on oak; **C. strobilinum**; II, III on evergreen oaks.

Rust, Eastern Gall. **Cronartium quercuum**, galls on trunk, branches, witches' brooms on two- and three-needle pines East to Rocky Mts. especially Southeast; II, III on oak, rarely chestnut.

Rust, Fusiform. **Cronartium fusiforme**, swellings in trunk, branches, in South; II, III on evergreen oaks; **Cronartium quercuum** f. sp. **fusiforme**, NC.

Rust, Lodgepole Pine Blister; Western Fusiform. **Cronartium coleosporioides** (*Peridermium harknessi*) (0, I), swellings on twigs, branches; trunk cankers; widespread in West; II, III on painted-cup, birdbeak, owl-clover, wood-betony; **C. stalactiforme** (0, I); II, III on Indian paintbrush.

Rust, Needle. **Coleosporium apocyanaceum** (0, I), Southeast; II, III on *Amsonia*; **C. asterum** (*C solidaginis*) (0, I), on all two-and three-needle pines in eastern U.S., western form CO to MT, WA; II, III on aster, goldenrod, and other composites; **C. campanulae** (0, I), Northeast to IN, NC; (I, III) on bellflower, loosestrife, Venus-looking-glass; **C. crowellii** (III), on pinon and limber pines, AZ, CO, NM; **C. delicatulum** (0, I), New England to FL and west to Great Plains; II, III on goldenrod; **C. elephantopodis** (0, I), NJ to FL, TX; II, III on *Elephantopus*; **C. helianthi** (0, I), NY to GA, OH; II, III on sunflower; **C. inconspicuum** (0, I), MD to GA, OH, TN; II, III on coreopsis; **C. ipomoeae** (0, I), NJ to FL, AZ, IL; II, III on morning-glory; **C. jonesii** (0, I), AZ, CO, NM; II, III on currant and gooseberry; **C. laciniariae**, NJ to FL; II, III on *Liatris*; **C. madiae** (0, I), CA, OR; II, III on composites; **C. minutum** (0, I), on loblolly and spruce pines, FL; II, III on *Forestiera*; **C. pinicola**, DE to NC, TN; **C. senecionis**, CO, RI; II, III on *Senecio*; **C. sonchiarvensis** (0, I), on Scotch pine, CT to NE; II, III on sow-thistle; **C. terebinthinaceae** (0, I), Southeast; II, III on *Parthenium* and *Silphium*; **C. vernoniae** (0, I), on two-and three-needle pines East and South; II, III on ironweed.

Rust, Pine-Pine Gall. **Endocronartium harknessii**, MN, MO, PA.

Rust, Piñon Blister. **Cronartium occidentale**.

Rust, Stem. **Cronartium appalachianum** (0, I), NC, TN, VA; II, III on *Buckleya*; **C. filamentosum; Peridermium weirii.**

Rust, Sweet-Fern Blister. **Cronartium comptoniae**, swellings on trunk and branches of two-and three-needle pines from northeastern to central and Great Lakes states; II, III on sweet-fern and sweet gale.

Rust, Western Gall. **Endocronartium harknessii**, VA, also twig necrosis and Witches' Broom, MA, ND.

Rust, White Pine Blister. **Cronartium ribicola** (0, I), swellings on trunk and branches of eastern white pine from New England to MN, NM, VA; on western white pine in Pacific Northwest; on sugar pine, CA; on limber pine, ND, SD, WY; II, III on gooseberry, currant, CA; on whitebark pine, CA, CO, NV; Mt. Rainier National Park and Crater Lake National Park.

SOOTY MOLD. **Fumago vagans**; **Capnodium pini**, widespread; **Scorias spongiosa**.

White pine blister rust is, of course, our foremost disease of pines, and full details are given under Rusts. Black currants are banned entirely in infected areas, red currants within 300 ft of pines. Of the other possible rusts on various pines, the only one seen in northern gardens is the aster rust (*Coleosporium asterum*), which is slightly disfiguring to the needles but not very damaging to general health. Fusiform rust is serious in the South. Brown needles may be due to one of the needle blights or needle cast fungi and also to winter drying. New shoots of Austrian pine turn brown from Diplodia tip blight, which should not be confused with the common discoloration caused by the pine shoot moth.

VIRUS. **Tomato Spotted Wilt**, GA (in *P. elliottii, P. taeda, P. palustris*).

WITCHES' BROOM. **Elytroderma deformans**, CA.

PISTACHIO (*Pistacia*)

BLIGHT, ROT ROOT. **Rhizoctonia solani**, CA.

BLIGHT, Shoot. **Botrytis cinerea**, CA; **Sclerotinia sclerotiorum**, CA; **Botryosphaeria rhodina**, CA; **Fusicoccum** (formerly *Botryosphaeria*).

BLIGHT, Thread. **Pellicularia koleroga**, FL, TX.

LEAF SPOT. **Phyllosticta lentisci**, TX; **Septoria pistaciarum**, AZ, NM, TX; **Alternaria** sp., NM.

LEAF SPOT, Brown. **Microascus cirrhosis**, Iran.

NEMATODE. **Meloidogyne** sp., CA; **Xiphinema index**, CA.

ROT, Root. **Phymatotrichum omnivorum**, AZ, CA, TX.

ROT, Sapwood. **Pleurotus ostreatus**, CA; **Schizophyllum commune**, CA.

WILT. **Verticillium albo-atrum**, CA.

PITHECELLOBIUM (Blackbead, Catsclaw)

BLIGHT, Twig. **Phomopsis** sp., FL.

LEAF SPOT. **Colletotrichum erythrinae**, TX; **Pestalotia funerea**, TX; **Phyllosticta pithecolobii**, TX.

NEMATODE. **Meloidogyne** sp.

RUST. **Ravenelia gracilis** (0, I, II, III), TX; **R. pithecolobii** (II, III); **R. siderocarpi** (II, III), TX.

PITTOSPORUM

BLIGHT, Southern; Wilt. **Sclerotium rolfsii**, FL, TX.

BLIGHT, Thread. **Pellicularia koleroga**, LA; **Rhizoctonia ramicola**, FL.

LEAF SPOT, Angular. **Cercospora pittospori**, FL, LA, SC to TX; **Phyllosticta** sp., AL; **Alternaria tenuissima**, FL.

NEMATODE, Root Knot. **Meloidogyne** sp.

ROT, Foot. **Diplodia** sp., FL.

ROT, Root. **Phymatotrichum omnivorum**, TX.

VIRUS. **Mosaic**, undetermined, CA; **Rough Bark**, undetermined, CA; **Variegation**. Variegated forms may be due to a virus.

WILT. **Verticillium albo-atrum**, CA.

PLANE-TREE, SYCAMORE (*Platanus*) (American Sycamore, *P. occidentalis*; California Sycamore, *P. racemosa*; London Plane, *P. acerifolia*; and Oriental Plane, *P. orientalis*)

ANTHRACNOSE; Leaf and Twig Blight. **Gnomonia platani** (*G. veneta, Gloeosporium platani, G. nervisequum*), general.

BACTERIAL Scorch. **Xylem limited Rickettsialike** bacteria, DC, LA, TX.

BLIGHT, Leaf. **Phloeospora multimaculans**, IN, TX.

BLIGHT, Twig; Canker. **Massaria platani**, CA, GA, IN, IA, KS, NJ.

CANKER. **Dothiorella** sp., NY; **Botryodiplodia theobromae**, MS; **Hypoxylon tinctor**, GA, LA, NC.

CANKER Stain; London Plane Blight. **Ceratocystis fimbriata** f. sp. **platani**, DE, MD, MS, MO, NJ, NY, NC, OH, PA, TN, VA, WV.

LEAF SPOT. **Cercospora platanicola** (*Mycosphaerella platanifolia*) NC to GA, IA, TX; spermatial stage is **Phyllosticta platani**; **Mycosphaerella stigmina-platani**, NC; **Septoria platanifolia**, GA, IA, MD, SC, TX, WV; **Stigmella platani-racemosae**, CA; **Cristutariella pyramidalis**, FL; **Tubakia dryina**, LA.

MISTLETOE. **Phoradendron serotinum (flavescens)**, AZ, NM, OK, TN, TX.

NONPARASITIC. **Rosy canker**. Illuminating gas in soil, MD, NJ, NY. **Decline**. Moisture deficiency, MS.

POWDERY MILDEW. **Microsphaera alni**, widespread; **Oidium obductum**, PA, VA, WV; **Phyllactinia corylea**, IN.

ROT, Heart, Trunk. **Fomes** spp.; **Steccherinum** (*Hydnum*) **erinaceus**, some-times on living trees.

ROT, Root. **Armillaria mellea**, MD, TX, WV; **Phymatotrichum omnivorum**, AZ, TX; **Phytophthora cinnamomi**, MD; **Clitocybe tabescens**, CA; **C. olearia**.

ROT, Wood. **Daedalea** spp., widespread; **Sterum** spp.; **Polyporus** spp.

The canker stain of London plane and American sycamore flared up in epidemic form around Philadelphia in 1935, killing thousands of street and ornamental trees from Newark to Baltimore before it was learned the fungus was spread in pruning and in tree paint as well as by certain beetles. Pruning is now restricted to winter months, and a disinfectant is added to the wound dressing. Sycamore anthracnose is serious in wet seasons, particularly when the mean daily temperature for 2 weeks after bud-break is below 55 °F.

PLUM (GARDEN), PRUNE (*Prunus domestica*)

BACTERIAL Crown Gall. **Agrobacterium tumefaciens**, widespread.

BACTERIAL Fire Blight. **Erwinia amylovora**, occasional, OR, WA.

BACTERIAL Leaf Spot; Black Spot. **Xanthomonas pruni**, eastern and southern states to WI, TX.

BACTERIAL Shoot Blight; Gummosis. **Pseudomonas syringae**, CA.

BLACK KNOT. **Dibotryon morbosum**, widespread except far West.

Blight, Blossom, Twig. **Monilinia laxa**, CA, OR, WA; **Botrytis cinerea**.
Blight, Thread. **Pellicularia koleroga**, LA.
Blight, Twig. **Diplodia** spp., secondary.
Blotch, Leaf. **Phyllosticta congesta**, GA, TX.
Blotch, Sooty; Flyspeck. **Zygophiala cryptogama, Z. wisconsinensis, Pseudocercosporella** sp., **Stomiopeltis** spp., IA.
Canker. **Phytophthora cactorum**, CA, IN; **Ceratocystis fimbriata**, CA; **Valsa** (*Cytospora*) **leucostoma**, dieback, widespread.
Leaf Curl; Witches Broom. **Taphrina** spp., occasional; **T. communis**, plum pockets; **T. pruni**, pockets, bladder plums; **T. insititiae**.
Leaf Scorch. **Xylella fastidiosa**, Southern CA.
Leaf Spot; Shot Hole. **Cercospora circumscissa**, CA, FL, MA, TX, WA; **Coccomyces prunophorae**, widespread; **Coryneum carpophilum**, CA to ID and WA; **Phyllosticta circumscissa**, IA, WA; **Septoria pruni**, TX.
Mistletoe. **Phoradendron serotinum** (**flavescens**), TX; **Viscum album**, CA, on *Prunus salicina* ("Santa Rosa" Plum).
Nematode, Lesion. **Pratylenchus pratensis**; **P. vulnus**.
Nematode, Root Knot. **Meloidogyne** spp., FL, TX.
Nonparasitic. **Brown Bark Spot**. Arsenical poisoning, MT.
Chlorosis. Alkaline soil, mineral deficiency, CA, FL.
Exanthema. Copper deficiency, CA, FL.
Gum Spot, Drought Spot. Irregular water supply, NY, Pacific Northwest.
Little Leaf. Zinc deficiency, CA, OR, WA.
Marginal Scorch. Fluorine injury, WA.
Myrobalan Asteroid Spot. Cause unknown, occasional.
Myrobalan Mottle. Genetic abnormality, occasional in seed stocks.
Wind Scab. Caused by developing fruit rubbing against other fruit, leaves and shoots during strong wind, CA.
Powdery Mildew. **Podosphaera oxyacanthae** var. **tridactyla**, occasional.
Pox, Plum. NY, PA.
Rot, Brown; Blossom Blight. **Monilinia fructicola**, general; **M. laxa**, MA, NY, RI.
Rot, Fruit. **Alternaria** sp., OR; **Botrytis cinerea**, CA, WA; **Cladosporium** sp., ID, OR; **Lambertella pruni**, CA.
Rot, Heart. **Fomes applanatus**, OR; **F. fulvus**, widespread; **Lenzites saepiaria**, Pacific Northwest; **Polyporus hirsutus**; **P. versicolor**, wide-spread.
Rot, Pink Mold. **Trichothecium roseum**, CA.
Rot, Root. **Armillaria mellea**, widespread; **Phymatotrichum omnivorum**, TX.
Rot, Silver Leaf. **Stereum purpureum**, WA.
Rust. **Tranzschelia discolor** (II, III), widespread, especially in South and Pacific Coast states; 0, I on anemone; **T. pruni-spinosae** (II, III); 0, I, on anemone, hepatica, thalictrum, buttercup.
Scab. **Cladosporium carpophilum**, widespread.
Virus. **Plum Line Pattern**; **Plum White Spot**; **Prune Diamond Canker; Prune Constricting Mosaic**; **Prune Dwarf**; **Cherry Vein Clearing**; **Little Peach**;

Peach Mosaic; **Peach Rosette**; **Peach Yellows**; **Ring Spot**; **Plum Rusty Blotch**; **Apricot Ring Pox**; **Stem Pitting; Cherry Green Ring Mottle,** North America.
WILT; Seedling Black Heart. **Verticillium albo-atrum**.

Plums are even more subject to brown rot than peaches and take the same spray schedule. This includes insect control to avoid infection through feeding injuries. Black knot is sometimes conspicuous on twigs, but diseased portions can be pruned out and a dormant lime-sulfur spray applied.

PLUM, WILD (*Prunus americana*)
BLOTCH Sooty. **Zygophiala jamaicensis**, IA.
FLYSPECK, Fruit. **Pseudocercosporella** spp.

PLUMERIA (*Frangipani*)
LEAF SPOT. **Cercospora plumeriae**, FL.
MISTLETOE. **Phoradendron serotinum** (**flavescens**), FL.
ROT, Root. **Phymatotrichum omnivorum**, TX.
RUST. **Coleosporium domingense** (II, III), FL; **C. plumeriae**, HI, LA.

PODOCARPUS
NEMATODE, Burrowing. **Rodopholus similis**, FL.
ROT, Root. **Clitocybe tabescens**, FL.

POINCIANA (*Caesalpinia*)
ANTHRACNOSE. **Gloeosporium** sp., FL.
BACTERIAL Crown Gall. **Agrobacterium tumefaciens**, FL.
ROT, Root. **Clitocybe tabescens**, FL; **Phymatotrichum omnivorum**, TX.
RUST. **Ravenelia humphreyana** (II, III), FL.

POMEGRANATE (*Punica granatum*)
ANTHRACNOSE; Fruit Spot. **Colletotrichum** sp., FL.
BLIGHT, Thread. **Pellicularia koleroga**, FL.
BLOTCH, Leaf; Fruit Spot. **Mycosphaerella lythracearum**, FL to MS, TX.
DECAY. **Pilidiella granati**, CA.
NEMATODE. **Meloidogyne** sp., FL.
ROT, Fruit. **Alternaria** sp., CA; **Aspergillus niger**, AZ, CA, TX; **Botrytis cinerea**, gray mold, cosmopolitan; **Nematospora coryli**, dry rot, CA; **Penicillium expansum**, blue mold, cosmopolitan; **Coniella granati**, NC.
ROT, Root. **Clitocybe tabescens**, FL; **Phymatotrichum omnivorum**, TX.
SPOT ANTHRACNOSE. **Sphaceloma punicae**, FL, LA, TX.

POND-SPICE (*Litsea*)
LEAF SPOT. **Cercospora olivacea**, GA.

POPLAR, ASPEN, COTTONWOOD (*Populus*)
BACTERIAL Limb Gall. **Agrobacterium tumefaciens**, CT, IA, MN, NE, TX.
BACTERIAL Wetwood. **Clavibacter humiferum**.
BLIGHT, LEAF. **Linospora tetraspora**, WA.

BLIGHT, Shoot. **Venturia populina** (*Didymosphaeria populina*), Northeast to WI; **V. tremulae**; **Colletotrichum gloeosporioides**, PA; **Glomerella cingulata**, WA, and leaf.

BRONZE LEAF DISEASE. **Apioplagiostoma populi**, Alberta, Canada.

CANKER, Branch and Trunk. **Dothichiza populea**, most serious on Italian varieties, widespread; **Hypoxylon pruinatum**, Southwest, Northeast and Great Lakes states, AZ, CA; **Nectria cinnabarina**; **N. galligena**; **Dothiora polyspora**, CO, NM, UT.

CANKER; DIEBACK. **Botryosphaeria ribis** var. **chromogena**, widespread, especially in the South; **Cytospora chrysosperma** (*Valsa sordida*); prevalent on ornamental poplars; **Valsa nivea**; **Ceratocystis tremullo-aurea**, Rocky Mts.; **C. fimbriata**, MN, PA; **Fusarium** (*Hypomyces*) **solani**, IA; **Phomopsis macrospora**, MS; **Ceratocystis** sp., Rocky Mts., MN, PA on aspen; **Septoria musiva**, Alberta and British Columbia, Canada on poplar.

CANKER, Sooty Bark. **Cenangium singulare**, Rocky Mts.

LEAF BLISTER, Yellow. **Taphrina aurea**, SC; **T. populina**, MS, New England to Great Lakes states; **T. johansonii**, catkin deformity, widespread; **T. populi-salicis**, CA, OR.

LEAF BLOTCH. **Septotinia populiperda**, ME, MD, MA, NY, VT.

LEAF BRONZE. **Apioplagiostoma populi**, Manitoba, Canada.

LEAF SPOT. **Marssonina** spp., widespread; **Mycosphaerella** (*Septoria*) **populicola**; **M. populorum** (*Septoria musiva*), also twig canker. **Phyllosticta** spp.; **Cercospora populina**, AL, LA, MO; **C. populicola**, TX; **Stigmina populi**, DE; **Marsonina brunnea**, WA and central U.S; **Septoria populicola**, WA.

LEAF SPOT; Ink Spot. **Ciborinia bifrons** (*Sclerotinia whetzelii*) in Northeast; **C. confundens** (*Sclerotinia bifrons*), CO to WY, OR, WA.

MISTLETOE. **Phoradendron serotinum (flavescens)**, AZ, NM, TX, **P. macrophyllum**

MISTLETOE, European. **Viscum album**, CA, on *Populus Fremontii* (Fremont Cottonwood) and *P. tremuloides* (Quaking Aspen)

NEMATODE, Dagger. **Xiphinema americanum**, SD, on cottonwood.

NEMATODE, Ring. **Criconemoides crotaloides**.

NONPARASITIC. **Chlorosis**. Iron deficiency, WY.

POWDERY MILDEW. **Uncinula salicis**, widespread; **Golovinomyces cichoracearum**, UT.

ROT, Heart. **Daedalea** spp., sometimes on living trees; **Fomes igniarius**, widespread; **F. pini**; **Pholiota adiposa** and **P. destruens**, New England states.

ROT, Root. **Armillaria mellea**, occasional; **Phymatotrichum omnivorum**, OK, TX; **Cylindrocladium scoparium**, of seedlings, GA.

ROT, Wood. **Collybia velutipes**, sometimes living trees, Rocky Mt. states; **Fomes** spp.; **Polyporus** spp.; **Lenzites saepiaria**, widespread.

ROT, Wound. **Schizophyllum commune**, cosmopolitan.

RUST, Leaf. **Melampsora bietis-canadensis** (II, III), New England to Great Plains; 0, I on hemlock; **M. albertensis** (II, III), CA, MT, NM; 0, I on Douglas-fir; **M. larici-populina** (II, III), WA, Quebec, Canada, Eastern North America;

M. medusae (II, III), through U.S. except far South; 0, I on larch; **M. populnea** (III), CO, RI, Pacific Coast; **M. occidentalis** (II, III), MT to CA, IA, WA, WI; 0, I unknown.
SPOT ANTHRACNOSE. **Sphaceloma populi**, OK.
VIRUS. **Poplar Decline**; **Poplar Mosaic**.

Cytospora canker is rather common on poplars lacking in vigor, but the Rio Grande cottonwood in the West is resistant. Avoid wounding, prune out twigs that have died back, and promote better growing conditions. Dothichiza canker may kill Lombardy poplars. Seriously diseased trees should be destroyed.

POTENTILLA (*Cinquefoil*)
DOWNY MILDEW. **Peronospora potentillae**, IA, NJ, TX.
LEAF SPOT. **Fabraea dehnii**, IA, NY, VT; **Marssonina potentillae**, CA, CO, MA, MI, NM, NY, WI; **Phyllosticta anserinae**, IL; **Ramularia arvensis**, WI.
POWDERY MILDEW. **Erysiphe polygoni**, CO; **Sphaerotheca macularis**, CO, WI.
RUST. **Phragmidium andersonii** (I, II, III), general.
VIRUS. **Tomato Ringspot,** NY.

POTHOS (*Epipremnum*)
ROT, Petiole, Leaf, Crown. **Phytophthora capsici**, MA.

POTHOS, IVY-ARUM (*Scindapsus*)
BACTERIAL Blight. **Pseudomonas cichorii**, FL.
NEMATODE, Burrowing. **Radopholus similis**, FL.
NEMATODE, Lesion. **Pratylenchus** sp.
NEMATODE, Root Knot. **Meloidogyne arenaria thamesi**; **M. incognita**.
ROT, Root. **Pythium splendens**, FL; **Rhizoctonia** sp., foot rot, MO.

PRICKLY-ASH (*Xanthoxylum americanum*)
CANKER, Stem. **Diplodia natalensis**, TX.
LEAF SPOT. **Cercospora xanthoxyli**, IN, TX; **Septoria pachyspora**, IA, NE.
POWDERY MILDEW. **Phyllactinia corylea**, widespread.
ROT, White Heart. **Fomes igniarius**.
RUST. **Puccinia andropogonis** var. **xanthoxyli** (0, I), IA, KS, MO, NE, TX, WI; II, III on *Andropogon*.

PRIVET (*Ligustrum*)
ANTHRACNOSE; Canker; Dieback. **Glomerella cingulata**, general, especially on common privet.
BACTERIAL Crown Gall. **Agrobacterium tumefaciens**, occasional.
BLIGHT, Leaf. **Ramularia** sp., NJ, WA.
BLIGHT, Thread. **Pellicularia koleroga**, FL.
DODDER. **Cuscuta** sp., FL.
GALL. **Phomopsis** sp., MD, TX; **P. ligustri-vulgaris**, blight, PA; **Phoma** sp., TX.
LEAF SPOT. **Cercospora adusta**, DE to AL, TX; **C. ligustri**, Gulf states; **Exosporium concentricum**, TX; **Phyllosticta ovalifolii**, MD, MS, TX; **Corynespora cassiicola**, FL.

LEAF SPOT, Algal. **Cephaleuros virescens**, Gulf states.
NEMATODE, Leaf. **Aphelenchoides fragariae**.
NEMATODE, Root Knot. **Meloidogyne** spp., southern states.
NONPARASITIC. **Chlorosis**. Manganese deficiency, FL.
PARASITIC Lichen. **Strigula elegans** and **S. complanata**, southern U.S., GA.
POWDERY MILDEW. **Nicrosphaeria alni**, IN, LA, NJ, OH, on *Ligustrum japonicum* (Wax-leaf Ligustrum).
ROT. **Rosellinia necatrix**, CA; **Ganoderma applanatus**, collar rot.
ROT, Root. **Armillaria mellea**, AR, CA, MS, TX; **Clitocybe monadelpha**, AR; **C. tabescens**, FL; **Phymatotrichum omnivorum**, AZ, OK, TX.
ROT, Wood. **Stereum hirsutum**, cosmopolitan; **Polyporus versicolor**, cosmopolitan.
SOOTY MOLD. Common after whiteflies in the South.
VIRUS. **Variegation**. Graft transmitted. **Chlorotic Spot,** LA; **Ligustrum Line Pattern**, AR.

PYRACANTHA (Firethorn)
BACTERIAL Fire Blight. **Erwinia amylovora**, widespread.
BLIGHT, Leaf. **Fabraea maculata**, LA.
BLIGHT, Silky Thread. **Rhizoctonia ramicola**, FL.
BLIGHT, Twig. **Diplodia crataegi**, PA.
CANKER, Felt Fungus. **Septobasidium cokeri** and **S. mariani**, on bark scales.
CANKER; DIEBACK. **Botryosphaeria ribis**, LA.
MISTLETOE, European. **Viscum album**, CA.
ROT, Root. **Armillaria mellea**, CA; **Phymatotrichum omnivorum**, TX.
SCAB. **Fusicladium pyracanthae**, widespread on leaves and fruit.
WILT. **Fusarium oxysporum**, FL.

Fire blight is the most common disease. It infects all species, but some are relatively resistant. Scab, after disfiguring on berries, can be prevented by spraying at bud-break and 10 and 20 days later.

Q

QUINCE (*Cydonia*)

ANTHRACNOSE, Northwestern. **Neofabraea malicorticis**, OR, WA.

BACTERIAL Crown Gall. **Agrobacterium tumefaciens**, general.

BACTERIAL Fire Blight. **Erwinia amylovora**, general.

BACTERIAL Hairy Root. **Agrobacterium rhizogenes**, ME to NC; Pacific Coast.

BLIGHT, Dothiorella Twig. **Botryosphaeria ribis**, TX.

BLIGHT, Leaf; Black Spot. **Fabraea maculata**, general.

BLIGHT, Thread. **Pellicularia koleroga**, NC, Gulf states.

CANKER, Perennial. **Neofabraea perennans**, OR.

CANKER, Trunk. **Nectria galligena**, OR; Twig, **Valsa leucostoma**.

FRUIT SPOT. **Leptothyrella pomi**, MO; **Mycosphaerella pomi**, New England to OH.

LEAF SPOT. **Phyllosticta** sp., DE.

NEMATODE, Root Knot. **Meloidogyne** sp., TX.

POWDERY MILDEW. **Phyllactinia corylea**, VA; **Podosphaera leucotricha**, CA, WA; **P. oxyacanthae**, NY, WV to IN.

ROT, Bitter; Canker. **Glomerella cingulata**, eastern and central states southward.

ROT, Black; Canker; Leaf Spot. **Physalospora obtusa**, eastern states to AL, TX.

ROT, Brown. **Monilinia fructicola**, eastern states, MS, OR, TX; **M. laxa**, also blossom and twig blight, Pacific Coast.

ROT, Fruit. **Alternaria mali**, IN; **Botrytis cinerea**, occasional; **Cephalothecium roseum**, pink mold, occasional; **Penicillium expansum**, cosmopolitan; **Phoma cydoniae**, pale rot, IL, MI; **P. mali**, IN.

RUST. **Gymnosporangium clavariiforme** (0, I), leaves, fruit, stems, CT, ME, NH; III on juniper; **G. clavipes** (0, I), orange rust, quince rust, on fruit, stems; III on juniper; **G. gracile** (0, I); III on juniper; **G. libocedri** (0, I), fruit, stems, OR; III on incense-cedar; **G. nelsonii** (0, I), leaves, stems, AZ, CO; III on juniper; **G. nidus-avis** (0, I), leaves, fruit, stems, CT, NY; III on juniper.

SCAB. **Venturia pirina**, CT.

SPOT ANTHRACNOSE. **Elsinoë piri**, WA.

R.K. Horst, *Field Manual of Diseases on Trees and Shrubs*,
DOI 10.1007/978-94-007-5980-0_20,

Quinces are subject to fire blight; infected branches should be cut out with the usual precautions. For leaf blight, spray when blossoms show pink, again when last of the petals are falling, and perhaps twice more at 2-week intervals. Brown rot is not very important on quince. To prevent rust, remove nearby susceptible junipers or spray them in spring as spore horns are developing on galls.

QUINCE, FLOWERING, JAPANESE, CHINESE (*Chaenomeles*)
BACTERIAL Crown Gall. **Agrobacterium tumefaciens**, occasional.
BACTERIAL Fire Blight. **Erwinia amylovora**, occasional.
BLIGHT, Leaf. **Fabraea macula**, AL, CT, NJ, NY.
CANKER, Felt Fungus. **Septobasidium burtii**, MS; **S. mariani**, NC.
CANKER, Twig Blight. **Botryosphaeria ribis**, AL, TX; **Phoma** sp., MD, TX; **Physalospora obtusa**, TX, eastern states.
FRUIT SPOT. **Mycosphaerella pomi**, IL, MD, also leaf blotch.
LEAF SPOT. **Cercospora cydoniae**, AL, GA.
NEMATODE, Root Knot. **Meloidogyne** sp.
ROT, Brown. **Monilinia fructicola**, leaf blight, MI; **M. laxa**, also blossom and twig blight, CA.
RUST, Quince. **Gymnosporangium clavipes** (0, I), on stems; CT, NJ; III on juniper; **G. libocedri** (0, I), on leaves; III on incense-cedar.

R

RABBITBRUSH (*Chrysothamnus*)

Leaf Spot. **Phloeospora bigeloviae**, CA.

Powdery Mildew. **Golovinomyces cichoracearum**, CA, MT, UT, WY; **Erysiphe polygoni** var. **sepulta**, CO, MT, UT, WY.

Rust. **Puccinia dioicae** (0, I), NM and CA; **P. grindeliae** (III), MT to NM, CA; **P. stipae** (0, I), MT to NM, CA; II, III on grasses.

Smut, Inflorescence. **Thecaphora pilulaeformis**, AZ.

RAMBUTAN (*Nephelium lappaceum*)

Canker, Stem. **Dolabra nepheliae**, HI, Puerto Rico.

Rot, Fruit. **Gliocephabotrichum simplex**, HI; **G. bulbilium**, HI; **Lasmenia** sp., HI, Puerto Rico; **Pestalotiopsis virgatula**, HI.

RASPBERRY (*Rubus*)

Anthracnose; Dieback; Gray Bark. **Gloeosporium allantosporum**, OR, WA.

Bacterial Blossom Blight. **Pseudomonas** sp., OR.

Bacterial Crown Gall. **Agrobacterium tumefaciens**, general; **A. rubi**, cane gall, NY, PA to IL, OR, WI.

Bacterial Fire Blight; Flower and Twig Blight. **Erwinia amylovora**, ME, NH, PA, WA, WI.

Bacterial Hairy Root. **Agrobacterium rhizogenes**, OR.

Bacterial, Witches' Broom. **MLO**, OR.

Blight, Cane. **Leptosphaeria coniothyrium**, general; **Physalospora obtusa**, IA, MD, MI, MO, ND; **Sclerotinia** sp., NY.

Blight, Spur. **Didymella applanata**, general.

Canker; Cane Spot. **Ascospora ruborum**, MA, OR, WI.

Canker; Dieback. **Glomerella cingulata**, AR, KY, MD, MI, MO, NJ, OH, RI, WV; **G. rubicola**, white bud, IL, NJ; **Botryosphaeria ribis**, FL; **Macrophoma rubi**, IL.

Dodder. **Cuscuta gronovii**, CT, IL, WI.

Downy Mildew. **Peronospora rubi**, WA.

Fruit Spot, Flyspeck. **Leptothyrium pomi**, IN, KY, MA.

R.K. Horst, *Field Manual of Diseases on Trees and Shrubs*,

DOI 10.1007/978-94-007-5980-0_21,

Leaf Spot. **Cylindrosporium rubi** (*Sphaerulina rubi*), also cane spot, common East of Rocky Mts.; **Mycosphaerella confusa** (*Cercospora rubi*), NJ to FL, IL, TX; **M. rubi**, general; **Pezizella oenotherae**, MD, MO, VA; **Septoria darrowi** (perhaps same as *Cylindrosporium rubi*) **Stigmatea rubicola**, black leaf and cane spot, MT, NM, NY, VT, WI.

Nematode, Lesion. **Pratylenchus vulnus.**

Nematode, Sheath. **Hemicycliophora** sp.

Nonparasitic. **Chlorosis.** Iron deficiency in West.

Powdery Mildew. **Phyllactinia corylea**, MI; **Sphaerotheca macularis**, Northeast, Northwest, CA, IL, MN.

Rosette, Double Blossom. **Cercosporella rubi**, IL, KY, MD, NY, PA.

Rot, Fruit. **Botrytis cinerea**, gray mold, cosmopolitan; **Phyllostictina carpogena**, MD; **Monilinia fructicola**, brown rot, IL; **Alternaria** sp., MA; **Rhizopus nigricans**, black mold, cosmopolitan.

Rot, Root. **Armillaria mellea**, OR, WA; **Rhizoctonia solani**, CO, ID, WA; **Xylaria** sp., WA; **Phytophthora erythroseptica**, Pacific Northwest; **P. rubi**, WA; **Pratylenchus penetrans**, WA.

Rust, Late. **Pucciniastrum americanum** (II, III), Northeast to ID, IL, NC, OH; 0, I on spruce.

Rust, Leaf. **Phragmidium rubiidaei** (0, I, II, III), northeastern and central states to CO; Pacific Northwest.

Rust, Orange. **Gymnoconia peckiana** (0, I, III), on black raspberry, Northeast to MN, Pacific Northwest; **Kunkelia nitens** (I), IL, IN, MI, OH.

Rust, Yellow. **Kuehneola uredinis** (0, I, II, III), IL, PA, DE to WI.

Spot Anthracnose. **Elsinoë veneta**, general but less common on red than black raspberry.

Virus. **Raspberry Leaf Curl; Red; Black Raspberry Necrosis,** TN; **Raspberry Streak**; **Raspberry Yellow Mosaic**, general; **Tomato Ring Spot,** cause of crumbly fruit, NY; **Tobacco Ring Spot,** NC; **Raspberry Bushy Dwarf**, OH; **Black Raspberry Witches' Broom Phytoplasma**, OR; **Crumbly Fruit Disease**, OR; **Rubus Yellow Net**, Alberta, Canada**; Blackberry Chlorotic Ringspot**, OR, WA.

Wilt. **Verticillium albo-atrum**, MA to NJ, OH, OR, WA.

Virus diseases are important on raspberries and cannot be controlled by spraying. Purchase healthy plants and set, if possible, 500 ft away from old patches. Inspect at least three times the first year, roguing all diseased plants, after first searing them with a blow torch or flame thrower so aphids will not carry the virus to nearby healthy bushes. Plants seldom recover from Verticillium wilt and never from orange rust, which is systemic. Crown gall is important on red raspberries; if infected plants are found, raspberries should not be replanted in the same soil for several years.

RATTAN VINE (*Berchemia*)

Rust. **Puccinia coronata** (0, I), VA to LA; II, III on oats and wild grasses.

REDBUD, JUDAS-TREE (*Cercis*)

CANKER; DIEBACK. **Botryosphaeria ribis** var. **chromogena**, DE, MD, NJ, NC, TX, VA.

DODDER. **Cuscuta exaltata**, TX.

DOWNY MILDEW. **Plasmopara cercidis**, TN.

LEAF SPOT. **Cercospora cercidis**, OK; **Cercosporella chionea**, IL, IN, KS, NC; **Mycosphaerella cercidicola**, general; **Phyllosticta cercidicola**, IN, WV; **Pestalotia guepini**, IL.

ROT, Root. **Phymatotrichum omnivorum**, TX; **Polyporus lucidus**, VA.

ROT, Wood. **Polyporus adustus**; **P. versicolor**; **Ganoderma lucidum** (*Polyporus lucidus*).

VIRUS. **Yellow Ringspot**, OR.

WILT. **Verticillium** sp., OH; of seedlings; **V. albo-atrum**, DC, VA; **V. dahliae**, DC; **Cylindrocladium** sp.

RED-BAY, SWAMP-BAY (*Persea borbonia*)

BLACK MILDEW. **Asterina delitescens**, VA to FL, TX; **Irenopsis martiniana**, AL, MS, TX; **Lembosia rugispora**, MS, NC; **Meliola amphitricha**, FL, MS; **Englerula carnea**, FL.

LEAF SPOT. **Cercospora purpurea**, GA to FL, MS; **Phyllosticta micropuncta**, MD to FL, TX; **Pestalotia** spp., general.

LEAF SPOT, Algal. **Cephaleuros virescens**, Gulf states.

ROT, Wood. **Polyporus hirsutus**; **P. mutabilis**.

WILT, Laurel. **Raffaelea albidum**, MS; **R. lauricola**, MS.

RHODODENDRON

BACTERIAL Crown Gall. **Agrobacterium tumefaciens**, OH.

BACTERIAL Leaf Spot. **Pseudomonas cichorii**, GA.

BLIGHT. **Phytophthora ramorum**, CA, NC, OR, WA; **P. tropicalis**, VA.

BLIGHT, Bud, Twig. **Briosia azalea**, GA, NJ, NY, NC, PA, TN, VA.

BLIGHT, Cutting. **Cylindrocladium scoparium** and **C. floridanum**, FL, OH.

BLIGHT, Flower Spot. **Ovulinia azaleae**, CA, SC.

BLIGHT, Gray. **Pestalotia macrotricha**, general after winter injury.

BLIGHT, Necrotic Shoots and Leaves. **Phytophthora irrigate** and **P. hydropathica**, MD, PA, VA, WV.

BLIGHT, Silky Thread. **Rhizoctonia ramicola**, FL.

BLIGHT,Twig. **Phomopsis** sp., CT, NJ, NY, OR; **Phytophthora hedraiandra**, PA.

CANKER; DIEBACK. **Botryosphaeria ribis**, MD, MA, NJ, NY; **Gloeosporium** sp., MD, NY; **Phytophthora cactorum**, MD, MA, NJ, NY, OH, PA, RI; **P. hedraiandra** (also leaf lesions), MN; **Glomerella cingulata**, MD; **Phytophthora cambivora**, NC.

DAMPING-OFF. **Alternaria** sp., CT; **Rhizoctonia solani**, CT, NJ, NY.

DODDER. **Cuscuta gronovii**, NJ.

GALL. **Exobasidium vaccinii**, leaf and shot, MA to FL, MS; **E. burtii**, yellow leaf spot, NJ; **E. vaccinii-uliginosae**, Witches' Broom, NJ.

Leaf Spot. **Cercospora handelii**, FL, NJ, NC, VA; **Coryneum rhododendri**, NC, PA, TN, VA; **Cryptostictis mariae**, KY, NY, TN, VA; **Discosia artocreas**, secondary, MD, NY; **Guignardia rhodorae**, CT, MD, MA, NJ, NY, PA, VA; **Gloeosporium ferrugineum**, NC; **Hendersonia concentrica**, NC, TX, WV; **Lophodermium schweinitzii**, NY to NC, TX; **Pestalopezia rhododendri**, TN, WV; **Mycosphaerella clintoniana**, NJ, NY, NC, OR, WA; **Phyllosticta rhododendri**, NJ; **Physalospora rhododendri**, PA, TN, VA; **Phomopsis rhododendri**, NJ; **Septoria rhododendri**, ME, NC; **S. solitaria**, NJ; **Venturia rhododendri**, MD, VA; **Phytophthora inflata**, OH; **P. insolita**, OH.

Leaf Spot, Tar. **Melasmia rhododendri**, AK.

Nonparasitic. **Chlorosis**. Mineral deficiency, usually iron, widespread.
Sunscald; Windburn. Severe winter injury in exposed locations.
Walnut Toxicity. Poisoning by root emanation from *Juglans nigra*.

Powdery Mildew. **Microsphaera alni**, MD, NJ, NY; **Erysiphe azalea** and **E. vaccinii**, Pacific Northwest.

Rot, Heart; Wood Rot. **Fomes annosus**, NC.

Rot, Root. **Armillaria mellea**, CA, NJ, NY; **Phymatotrichum omnivorum**, TX.

Rust. **Chrysomyxa ledi** var. **rhododendri**, CA, WA; **C. roahnensis** (II, III), NC, TN; 0, I on spruce; **C. piperiana** (II, III), CA, OR, WA; **Pucciniastrum vaccinii** (II), CT, NJ, RI.

Spot Anthracnose. **Sphaceloma** sp., WA.

Virus. **Rhododendron Necrotic Ringspot,** OR.

Wilt. **Phytophthora cambivora**, NC; **P. cinnamomi**, MD, NJ, NY, PA, OH, VA; **P. citricola**, OH, VA; **P. cactorum**, VA; **P. citrophthora**, VA; **P. nicotianae**, VA; **P. tropicalis**, VA; **P. hibernalis**.

Most rhododendron leaf spots are not worth worrying about. Some come after winter injury; some are definitely parasitic but not serious. Winter and early spring sun will turn some of the foliage brown. Do not prune out supposedly dead twigs and branches too soon; wait for new growth to start. An accumulation of matted wet leaves around the trunk fosters root and collar rot. Be cautious in the use of aluminum sulfate to acidify soil; sulfur is somewhat safer.

RIBBON-BUSH (*Homalocladium*)
Powdery Mildew. **Erysiphe polygoni**, NY, PA, WI.

RICE-PAPER PLANT (*Tetrapanax*)
Nematode, Root Knot. **Meloidogyne** sp., FL.

ROCK-ROSE (*Cistus*)
Rot, Root. **Armillaria mellea**.

ROLLINIA
Dieback, Fruit Rot. **Glomerella cingulata**, FL.

ROSE (*Rosa*)
Bacterial Blast. **Pseudomonas syringae**.
Bacterial Crown Gall. **Agrobacterium tumefaciens**, general.

BACTERIAL Hairy Root. **Agrobacterium rhizogenes**, MD, PA, TX, VA, New England.

BLACKSPOT. **Diplocarpon rosae**, general; **Marssonina rosae**, MS.

BLIGHT, Blossom. **Botrytis cinerea**, cosmopolitan; gray mold on canes in storage; **Dothiorella** sp., LA, VA.

BLIGHT, Cane. **Physalospora obtusa**, CT, to AL, KS, TX; **Gloeosporium** spp., widespread.

BLIGHT, Southern. **Sclerotium rolfsii**, FL, KS, TX.

BLIGHT, Thread. **Pellicularia koleroga**, FL, LA.

CANKER, Brand. **Coniothyrium wernsdorffiae**, MN, NY, PA (also reported from CO, IN, MS, TX, but probably mistaken for *C. fuckelii*).

CANKER, Brown. **Cryptosporella umbrina**, MA to FL, CA, ID, MI, NE, TX.

CANKER, Common, Graft. **Leptosphaeria coniothyrium** (*Coniothyrium fuckelii*), general; Graft, **C. rosarum**, CA, IA, MA, MN, NJ, PA, TX.

CANKER, Crown. **Cylindrocladium scoparium**, in greenhouses, MA to GA, IL, TX.

CANKER; DIEBACK. **Botryosphaeria ribis** var. **chromogena**, AL, MD, TX, VA; **Cryptosporium minimum**, OR, PA; **Diplodia** spp., probably secondary; **Griphosphaeria corticola** (*Coryneum microstictum*), NH to AL, ND; Pacific Northwest; **Nectria cinnabarina**, coral spot, MA to VA, AK, WA; **Cytospora** sp., KY, PA, VA, WA; **Didymella sepincoliformis**, MD; **Glomerella cingulata**, MD, NJ, VA; **Macrophoma** sp., TX, VA; **Botryodiplodia theobromae**, MN; **Trichothecium roseum**, MN.

DODDER. **Cuscuta indecora** and **C. paradoxa**, FL, TX.

DOWNY MILDEW. **Peronospora sparsa**, mostly under glass, ME to FL, CA, IA; reported outdoors in DE.

FOLIAR LESIONS, Black. **Xanthomonas perforans**, FL.

LEAF SPOT. **Alternaria** sp., VA to AL, TX; **A. brassicae** var. **microspora**, TX; **Cercospora puderi**, FL, GA; **Mycosphaerella rosicola** (*Cercospora rosicola*), general; **M. rosigena**, doubtfully distinct from *M. rosicola*, reported from South; **Monochaetia compta**, AK, IA, KS, MD; **Pezizella oenotherae**, also cane spot, NJ to FL, MI, TX; **Phyllosticta rosae**, NY to FL, IN; **P. rosae-setigerae**, IN; **Septoria rosae**, MS, NJ, SC.

MOLD, Black, of Grafts. **Chalaropsis thielavioides**, CA, IL, NY, PA on understock from OR and WA.

MOLD, Leaf and Bud. **Cladosporium** sp. and **C. fuscum**, AK, CA, MD, MN, MS, OK, TX.

NEMATODE, Dagger. **Xiphinema diversicaudatum**, greenhouses in Northeast; **X. americanum**; **X. krugi**.

NEMATODE, Leaf. **Aphelenchoides** spp.; Lance, **Hoplolaimus** sp.

NEMATODE, Lesion. **Pratylenchus pratensis**; **P. scribneri**; **P. vulnus**.

NEMATODE, Ring. **Criconemoides** sp.; Pin, **Paratylenchus** spp.

NEMATODE, Root Knot. **Meloidogyne** sp.; **M. hapla**.

NEMATODE, Spiral. **Helicotylenchus** spp.; Sheath, **Hemicycliophora** spp.

NEMATODE, Sting. **Belonolaimus gracilis**; Stubby Root, **Trichodorus** spp.

NEMATODE, Stylet. **Tylenchorhynchus** spp.

NONPARASITIC. **Boron Deficiency**. Leaves distorted, greenhouse.

Chlorosis. Iron deficiency, upper leaves yellow, with green veins; nitrogen deficiency, lower leaves pale; potassium deficiency, leaves grayish, may drop, stems weak.

Leaf Scorch. Marginal, potash deficiency. In greenhouses scorch may be boron and calcium deficiency.

Mercury Toxicity. In greenhouses when paint containing mercury used on sash.

Pedicel Necrosis. Collapse of flower stem, cause unknown.

Weed-killer Injury. Leaves fernlike, twisted when 2,4-D used in vicinity.

PETAL SPOT. **Bipolaris** (*Helminthosporium*) **setariae**, FL.

POWDERY MILDEW. **Podosphaera pannosa**, general; **S. macularis**, not readily distinguished from *S. pannosa*; **Phyllactinia corylea**, WA.

ROT, Dry. **Fusarium oxysporum**, Brazil.

ROT, Root. **Armillaria mellea**, CA, MS, OR, TX, WA; **Clitocybc tabescens**, FL; **Fusarium** spp., occasional, especially in the South; **Phymatotrichum omnivorum**, AZ, TX; **Ramularia macrospora**, MD.

RUST. **Phragmidium americanum** (0, I, II, III), on leaves of cultivated and native roses, ME to NC, ND, TX; **P. montivagum** (0, I, II, III), on native species, SD to NM, AZ, WA; **P. mucronatum** (*P. disciflorum*), the common rust of cultivated roses, possibly on native species (0, I), on leaves and stems; II, III on leaves, eastern states to Rocky Mts.; Pacific Coast; **P. fusiforme** (*P. rosae-acicularis*) (0, I, II, III), on native species, MI to CO, CA, WY; **P. rosae-californicae** (0, I, II, III), on natives, AZ, CA, MT, OR; **P. rosae-pimpinellifoliae** (*P. subcorticium*) (0, I), on stems; II, III on leaves of brier and sweetbrier groups, northern U.S.; **P. rosicola** (III), CO, MT, NE, on native spp.; **P. speciosum** (0, I), on leaves and stems, III on stems of cultivated and native roses, general except far South.

SPOT ANTHRACNOSE. **Elsinoë rosarum**, ME to FL, KS, MI, MO, TX; Pacific Coast.

VIRUS. **Rose Mosaic** (in part **Prunus Necrotic Ring Spot** and **Apple Mosaic**), Pacific Coast and eastern states in greenhouses and in gardens on plants shipped from the West Coast; **Rose Rosette** (Witches' Broom), CA, KS, MD, MO, OR; **Rose Streak**, MD, NJ, NY, TX, VA; **Rose Yellow Mosaic**, CA, IL, MD, NY, PA, VA; **Crinkle**. On Manetti understock, sometimes garden roses, Pacific Coast, MD, NY, PA, TX, VA; **Rose Leaf Curl**, CA; **Rose Spring Dwarf**, CA.

UNKNOWN. **Speckle**. Chlorotic flecks in leaves, not transmitted by grafting, MD, NJ, NY, PA, TX, VA.

WILT. **Verticillium albo-atrum**, AR, CA, IL, NJ, NY; probably wide spread.

Blackspot, brown canker, powdery mildew, and rust are the big four diseases of garden roses.

Blackspot is almost inevitable except in some dry western states, and shows up even there when overhead watering is substituted for the usual irrigation. It can be controlled by regular weekly spraying or dusting. Powdery mildew, a problem on the Pacific Coast, is increasing in eastern gardens. Brown canker and other cane diseases are best controlled at spring pruning, by cutting out infected canes and cutting other canes just above a bud, not leaving any stub to die back. Cankers are increased by excessive winter protection. Where temperatures permit, as in the Central Atlantic region, eliminate soil mounding and other special winter treatment. Roses are sensitive to many chemicals; it is important to distinguish spray injury from blackspot and not increase the dosage because you think you are not getting control. Combination sprays or dusts should take care of most diseases, as well as insects, in one operation. The bacterial crown gall is occasionally present on plants purchased from a nursery. Ask for a replacement; do not contaminate your soil by planting such a bush.

ROSE – ACACIA (*Robinia hispida*)
LEAF SPOT. **Alternaria fasciculata**, ND.
ROT, Root. **Phymatotrichum omnivorum**, TX.

ROSELLE (*Hibiscus sabdariffa*)
ANTHRACNOSE, Pod Spot. **Colletotrichum gloeosporioides**, FL.
BLIGHT, Gray Mold. **Botrytis cinerea**, MD.
BLIGHT, Southern. **Sclerotium rolfsii**, TX.
LEAF SPOT. **Cercospora hibisci**, TX.
NEMATODE, Root Knot. **Meloidogyne** sp., TX.
POWDERY MILDEW. **Microsphaera euphorbiae**, AL, FL.
ROT, Fruit, Stem. **Fusarium** sp., FL; **Phytophthora parasitica**, TX.
ROT, Root. **Phymatotrichum omnivorum**, TX; **Rhizoctonia solani**, TX.
WILT. **Fusarium oxysporum**, FL.

ROSEMARY (*Rosmarinus*)
BLIGHT, AERIAL, ROT ROOT. **Rhizoctonia solani**, LA, OK.
ROT, Root. **Phymatotrichum omnivorum**, TX.

ROSE-OF-SHARON, SHRUB-ALTHAEA (*Hibiscus syriacus*)
LEAF SPOT. **Cercospora malayensis**, GA; **Phyllosticta hibiscina**, OK; **P. syriaca**, NY.
NEMATODE, Root Knot. **Meloidogyne** sp., MS, TX.
ROT, Root. **Phymatotrichum omnivorum**, TX.
RUST. **Kuehneola malvicola** (II, III), Gulf states.

RUBBER-PLANT (*Ficus elastica*)
ANTHRACNOSE. **Glomerella cingulata**, general; **Gloeosporium** sp., WA.
BACTERIAL Crown Gall. **Agrobacterium tumefaciens**, CA, TX.
CANKER; DIEBACK. **Physalospora rhodina**, GA.
LEAF SPOT. **Alternaria** sp., IN, OH, TX; **Leptostromella elastica**, NY, Northeastern states; **Mycosphaerella bolleana**, GA; **Phyllosticta** sp., MD, NY; **Phyllosticta roberti**, Gulf states, MD, NY; **Stemphylium elasticae**, secondary; **Trabutia** (*Phyllachora*) **ficuum**, black spot, FL.
NEMATODE, Leaf. **Aphelenchoides besseyi**, FL.
NEMATODE, Root Knot. **Meloidogyne** spp.; **M. incognita**.
NEMATODE, Spiral. **Helicotylenchus nannus**.

RUSSIAN-OLIVE, SILVERBERRY (*Elaeagnus*)
CANKER. **Fusicoccum elaeagni**, IL, MO.

S

SAGE (*Salvia*) (Includes Blue, Clary, Sauceleaf, Scarlet, Black Ornamental Forms)
Bacterial, MLO. **California Aster Yellows**, CA.
Blight, Southern. **Sclerotium rolfsii**, IL.
Damping-Off. **Pythium debaryanum**, OH; **Rhizoctonia solani**, CT, IL, NJ, OH.
Downy Mildew. **Peronospora lamii**, IA, LA; **P. swinglei**, KS.
Leaf Spot. **Cercospora salviicola**, OK, TX; **Ramularia salviicola**, OK; **Myrothecium roridum**, FL.
Nematode, Leaf. **Aphelenchoides fragariae**, DE, NJ.
Nematode, Root Knot. **Meloidogyne** spp., AZ, NJ; **M. javanica**.
Powdery Mildew. **Golovinomyces cichoracearum**, CA.
Rot, Charcoal. **Macrophomina phaseoli**, SC.
Rot, Root. **Fusarium oxysporum**, NC; **Phytophthora dreschleri**; **P. parasitica**, NC; **Phymatotrichum omnivorum**, TX; Stem, **Sphaeropsis salviae**, MS.
Rot, Root and Crown. **Phytophthora cryptogea**, CA.
Rust. **Aecidium subsimulans**, AZ; **Puccinia ballotaeflorae** (II, III), TX; 0, I unknown; **P. caulicola** (0, I, II, III), IA to TX, NM; **P. farinacea** (0, I, II, III), AL, AZ, KS, MS, MO, NE, OK, TX; **P. salviicola** (0, I, II, III), FL, TX; **P. vertisepta** (0, I, III), AZ, NM; **P. melliflora** (I, III), CA.
Stem Spot, Lesion; Wilt. **Sclerotinia sclerotiorum**, ND.
Virus. **Tomato Spotted Wilt**.

SAGE-BRUSH (*Artemisia*)
Blight, Gray Mold. **Botrytis cinerea**, AK.
Blight, Stem. **Sclerotium** sp., OR.
Canker, Stem Gall, Black Knot. **Syncarpella tumefaciens**, CA, MT, NV.
Dodder. **Cuscuta** sp., TX.
Downy Mildew. **Peronospora leptosperma**, CA, IA, KS, ND, WI.
Leaf Spot. **Cercospora ferruginea**, WI; **C. olivacea**, NY; **Cylindrosporium artemisiae**, WA, WI; **Heterosporium** sp., AK; **Phyllosticta raui**, MT; **Ramularia artemisiae**, NY, WI; **Septoria artemisiae**, WA.

R.K. Horst, *Field Manual of Diseases on Trees and Shrubs*,
DOI 10.1007/978-94-007-5980-0_22, © Springer Science+Business Media Dordrecht 2013

Nematode, Root Knot. **Meloidogyne** sp., AL.
Powdery Mildew. **Golovinomyces cichoracearum**, WI to NM, CA, WA.
Rot, Root. **Phymatotrichum omnivorum**, TX.
Rust. **Puccinia atrofusca**, IA to TX, CA, OR; II, III on *Carex*; **P. millefolii** (III), ND to TX, AK, CA, WA; **P. tanaceti** (0, I, II, III), WI to TX, CA, WA; 0, I, II, III on **Tanacetum** and II on chrysanthemum; **Uromyces oblongisporus** (III), WY.

SAINFOIN (*Onobrychis*)
Damping-Off. **Rhizoctonia solani**, TX; **Sclerotium rolfsii**, TX.
Nematode, Root Knot. **Meloidogyne hapla**, WY.
Rot, Stem. **Rhizoctonia solani**, TX; **Sclerotium rolfsii**, TX.

ST. ANDREWS CROSS, ST. PETERSWORT (*Ascyrum*)
Leaf Spot. **Cladosporium gloeosporioides**, AL.
Rust. **Uromyces triquetrus** (0, I, II, III), MS, NJ, TX.

SALAL (*Gaultheria shallon*)
Anthracnose. **Colletotrichum acutatum**, British Columbia, Canada.
Black Mildew. **Meliola** sp.
Leaf Spot. **Dasyscyphus gaultheriae**, CA, OR, WA; **Mycosphaerella gaultheriae**, Pacific Coast, AK; **Pestalopezia brunneo-pruinosa**, CA, OR, WA; **Phyllosticta gaultheriae**, general.
Leaf, Stem Lesions. **Phoma exigua**, British Columbia, Canada.
Powdery Mildew. **Microsphaera alni**, OR.
Spot Anthracnose. **Elsinoë ledi**, OR, WA.

SALT BUSH (*Atriplex*)
Downy Mildew. **Peronospora farinosa**, MT, TX.
Gall, Leaf, Stem. **Urophlyctis pulposa**, ND, TX.
Leaf Spot. **Cercospora dubia**, widespread; **Stagonospora atriplicis**, KS, NE, NJ, NY, PA.
Nematode, Root Knot. **Meloidogyne** sp., CA; root gall, **Heterodera schachtii**, UT.
Rust. **Puccinia aristidae** (0, I); II, III on grasses; **Uromyces shearianus** (0, I, III).
Virus. **Beet Curly Top**; **Iris Yellow Spot**, UT.

SANCHEZIA
Rot, Mushroom Root. **Clitocybe tabescens**, FL.

SAND-MYRTLE (*Leiophyllum*)
Gall, Leaf. **Exobasidium vaccinii**, NJ, NC.

SAPODILLA (*Achras*)
Gall, Limb. **Pestalotia scirrofaciens**, FL, TX.
Leaf Spot. **Phyllosticta** sp., FL; **Septoria** sp., FL.
Rot, Root. **Phymatotrichum omnivorum**, TX.
Rust. **Scopella** (*Uredo*) **sapotae** (II), FL.
Spot Anthracnose. **Elsinoë lepagei**, FL.

SASSAFRAS

CANKER, Branch, Trunk. **Nectria** sp., CT to WV, MS.

CANKER; DIEBACK. **Physalospora obtusa**, NY to GA.

LEAF SPOT. **Septoria** sp., NY; **Actinothyrium gloeosporioides** (*Actinopella dryina*); **Cristulariella pyramidalis**; **Phyllosticta illinoiensis**, IL, MA; **P. sassafras**, NY to GA, TX, IL.

MISTLETOE. **Phoradendron serotinum** (**flavescens**), TX.

POWDERY MILDEW. **Phyllactinia corylea**, MI.

ROT, Heart, Trunk. **Daedalea confragosa**, IN, NY; **Fomes igniarius**, OH, VA; **F. ribis**, MO.

ROT, Root. **Armillaria mellea**, PA; **Phymatotrichum omnivorum**, TX.

ROT, Wood. **Daldinia vernicosa**; **Hymenochaete agglutinans**; **Hypoxylon** spp.; **Polyporus** spp., sometimes on living trees; **Poria ferruginosa**; **Schizophyllum commune**, NY; **Trametes sepium**, IN.

VIRUS. **Mosaic**, NY; **Yellows**, TX, unidentified.

WILT, Laurel. **Raffaelea lauricola**, FL, MS, SC.

The undetermined yellows disease causes fasciation of tops, leafroll, and dwarfing of leaves.

SEA-GRAPE, DOVE-PLUM (*Coccoloba*)

SPOT, Tar. **Phyllachora simplex**, FL.

SENECIO (Groundsel)

BACTERIAL, MLO. **California Aster Yellows**.

GALL, Leaf. **Synchytrium aureum**, WI.

LEAF SPOT. **Cercospora senecionicola**, WI; **C. senecionis**, TX; **Gloeosporium senecionis**, CA; **Phyllosticta garrettii**, OR, UT, WY; **Ramularia filaris**, CO, MT; **R. pruinosa**, CO, WY; **R. senecionis**, CA, CO; **Septoria cacaliae**, AL, IN, TX; **S. senecionis**, CA.

NEMATODE, Leaf. **Aphelenchoides ritzemabosi**.

POWDERY MILDEW. **Golovinomyces cichoracearum**, ID, MN, NE, VA, WA; **Sphaerotheca macularis**, CO, MT, UT, WY.

ROT, Root. **Phymatotrichum omnivorum**, TX; **Rhizoctonia solani**, IL, NJ.

ROT, Stem. **Phytophthora** sp., NJ; **Sclerotinia sclerotiorum**, LA.

RUST. **Coleosporium occidentale** (II, III), CA, CO, ID, MT, OR, WA, WY; 0, I unknown; **C. senecionis** (II, III), CO, RI; **Puccinia angustata** var. **eriophori** (0, I), CT, IA, MN, NH, UT, VT; II, III on *Eriophorum* and *Scirpus*; **P. expansa**, CA, UT, WA, WY; **P. lagenophorae**, CA; **P. recedens**, CO, CT to NC, IA, TN, ND to OR, WA, WY; **P. dioicae** (0, I), NE, NM, TX; II, III on *Carex*; **P. stipae** (0, I), CO, NE, WY; II, III on *Stipa*; **P. subcircinata** (0, I, III), ID, NE, NV, NM, ND, UT, WA; **Baeodromus californicus** (III), CA.

SMUT, White. **Entyloma compositarum**, KS, MD, NE, PA, TX, WI.

VIRUS. **Tomato Spotted Wilt**.

WHITE RUST. **Albugo tragopogonis**, CA, CO, IN, MO, MT, NE, UT, WA.

WILT. **Fusarium** sp., NJ; **Verticillium albo-atrum**, WA.

SEQUOIA (Redwood and Giant Sequoia)

Blight, Needle. **Chloroscypha chloromela**; **Cercospora sequoiae**, MD, PA; **Mycosphaerella sequoiae**; **Pestalotia funerea**, TX; **Phytophthora citrophthora**, also branch dieback, CA.

Blight, Seedling. **Botrytis douglasii**, CA, OH, PA.

Blight, Twig. **Phomopsis juniperovora**; **Botrytis cinerea**.

Burls, Gall, on trunk. Cause unknown.

Canker. **Botryosphaeria dothidea**, VA. [Redwood, Dawn] [U.S. Nat'l Arboretum, Washington, DC]; **B. dothidea** (= **B. ribis**), CA; **Phytophthora ramorum** (branch and leaf discolor), CA.

Canker, Bark. **Dermatea livida**.

Nematode. **Meloidogyne hapla**, **M. incognita**, **M. javanica**, **Pratylenchus penetrans**, **P. vulnus**, **Xiphinema bakeri**, CA.

Rot, Charcoal. **Macrophomina phaseoli**.

Rot, Root. **Armillaria mellea**; **Phymatotrichum omnivorum**, TX.

Rot, Trunk; Heart. **Fomes annosus**; **Ganoderma sequoiae**; **Poria sequoiae**; **P. albipellucida**.

Rot, Wood. **Hymenochaete tabacina**; **Lenzites saepiaria**; **Merulius hexagonoides**; **Polyporus** spp.; **Schizophyllum commune**; **Stereum** spp.; **Trametes** spp.

SHINING WILLOW (*Salix lucida*)

Leaf Spot. **Septoria musiva**, Quebec, Canada.

Bacterial, Yellows. **Spiroplasma citri**, IL.

SILK-TASSEL BUSH (*Garrya*)

Black Mildew. **Lembosia lucens**, CA.

Leaf Spot. **Cercospora garryae**, CA, TX; **Dothichiza garryae**, CA; **Phyllosticta garryae**, CA, TX.

Rot, Root. **Phymatotrichum omnivorum**, TX.

SKIMMIA

Virus. **Tobacco Ring Spot**, NY.

SMITHIANA

Virus. **Smithiana Potexvirus.**

SMOKE-TREE (*Cotinus*)

Leaf Spot. **Cercospora rhoina**, AL; **Pezizella oenotherae**, MD; **Septoria rhoina**, CT, MA, NY, VA; **Gloeosporium** sp., IL.

Rot, Root. **Phymatotrichum omnivorum**, TX.

Rust. **Pileolaria cotini-coggyriae**, GA, RI.

Wilt. **Verticillium albo-atrum**, CT, IL, NE, NJ, NY.

SMOKE-TREE (*Dalea*)

Leaf Spot. **Cercospora daleae**, KS.

Mistletoe. **Phoradendron californicum**, CA to TX.

Rot, Root. **Phymatotrichum omnivorum**, TX.

RUST. **Pileolaria cotini-coggyriae**, AR; **Puccinia andropogonis** (0, I), SD to KS; II, III on *Andropogon*; **P. paroselae** (II, III), CA.

SNOWBERRY (*Symphoricarpos*)

ANTHRACNOSE; Black Berry Rot; Twig Canker. **Glomerella cingulata**, widespread, MA to VA, IL and WI.

BACTERIAL Crown Gall. **Agrobacterium tumefaciens**, MD.

BACTERIAL Hairy Root. **Agrobacterium rhizogenes**, IA.

LEAF SPOT. **Ascochyta symphoricarpophila**, NY; **Cercospora symphoricarpi**, MT, SD; **Phyllosticta symphoricarpi**, NM, NY, WA; **Lasiobotrys symphoricarpi**, black spot, CO, UT, WY; **Septoria oedospora**, CO; **S. signalensis**, WY; **S. symphoricarpi**, ND to CO, CA and WA.

POWDERY MILDEW. **Microsphaera diffusa**, general; **Podosphaera oxya-canthae**, WA.

ROT, Berry. **Alternaria** sp., CO, CT, MA, NY; **Botrytis cinerea**, CT, MA, NY.

ROT, Collar. **Fomes ribis**, KS, MT.

ROT, Root. **Helicobasidium purpureum**, TX.

RUST. **Puccinia crandallii** (0, I), CA, CO, ID, MT, ND, UT, WA, WY; II, III on grasses; **P. symphoricarpi** (III), MT to CA, AK.

SPOT ANTHRACNOSE; Scab. **Sphaceloma symphoricarpi**, ME to VA, AR, CA, CO, OR, WA, WI.

Anthracnose and spot anthracnose, scab, commonly disfigure berries.

SOAPBERRY, SOUTHERN (*Sapindus saponaria*)

BLIGHT, Thread. **Pellicularia koleroga**, FL.

LEAF SPOT; Dieback. **Glomerella cingulata**, FL; **Phyllosticta sapindii**, FL.

SOAPBERRY, WESTERN (*Sapindus drummondii*)

BLIGHT, Leaf. **Cylindrosporium griseum**, OK, TX.

LEAF SPOT; Dieback. **Glomerella cingulata**, TX; **Mycosphaerella sapindii**, MO.

MISTLETOE. **Phoradendron serotinum** (**flavescens**), AZ, NM, TX.

POWDERY MILDEW. **Uncinula circinata**, TX.

ROT, Root. **Helicobasidium purpureum**, TX.

VIRUS. **Mosaic**, TX.

This is one of the few plants reported resistant to Texas root rot.

SOPHORA (Pagoda Tree, Silky Sophora, Mescalbean)

BLIGHT, Twig. **Nectria cinnabarina**, CT, NY.

CANKER. **Fusarium lateritium**, NJ.

DAMPING-OFF. **Rhizoctonia solani**, CT.

DIEBACK. **Diplodia sophorae**, OH.

LEAF SPOT. **Phyllosticta sophorae**, OK, TX.

MISTLETOE. **Phoradendron serotinum** (**flavescens**), TX.

NEMATODE, Root Knot. **Meloidogyne** sp., MD.

POWDERY MILDEW. **Microsphaera alni**, CT.

Rot, Root. **Phymatotrichum omnivorum**, TX.
Rust. **Uromyces hyalinus** (0, I, II, III), SD to TX, AZ, WY.
Virus. **Brooming Disease**. Unidentified.

SPHACELE (Pitcher-Sage)
Rust. **Uredo sphacelicola** (II), CA.

SPICE-BUSH (*Lindera*)
Canker. **Botryosphaeria ribis** var. **chromogena**, MD.
Leaf Spot. **Phyllosticta linderae**, DE, IN, WV; **P. lindericola**, WV.
Mistletoe. **Phoradendron serotinum (flavescens)**, eastern states.
Rot, Root. **Phymatotrichum omnivorum**, TX.

SPIRAEA (Native Hardhack, Meadowsweet)
Blight, Seedling; Stem Girdle. **Thelephora terrestris**, ID.
Canker. **Cryptodiaporthe macounii**, NY.
Leaf Spot. **Cercospora rubigo**, CA, KS, OR, WI; **Cylindrosporium** spp., ID, NY; **Phloeospora salicifoliae**, NY to KS, TX, WA.
Powdery Mildew. **Podosphaera oxyacanthae**, widespread; **Sphaerotheca macularis**, CT, MI, NY, PA.

SPIREA, ORIENTAL FLOWERING (*Spiraea*)
Bacterial Fire Blight. **Erwinia amylovora**, MD, NJ, NC, VA.
Bacterial Hairy Root. **Agrobacterium rhizogenes**, IA.
Leaf Spot. **Cylindrosporium filipendulae**, IA, OR, WA, WI.
Nematode, Root Knot. **Meloidogyne** spp., FL, MS; **M. hapla**.
Powdery Mildew. **Microsphaera alni**, CT; **Podosphaera oxyacanthae**, widespread; **Sphaerotheca macularis**, TX.
Rot, Root. **Phymatotrichum omnivorum**, AZ, TX.

SPONDIAS (Mombin)
Nematode, Root Knot. **Meloidogyne** sp., FL.
Spot Anthracnose; Mombin Scab. **Sphaceloma spondiadis**, FL.

SPRUCE (*Picea*)
Blight, Brown Felt. **Herpotrichia nigra**, MT, Northern Rocky Mts., Pacific Northwest; **Neopeckia coulteri**.
Blight, Needle. **Sirula macrospora**, ND; **Stigmina lautii**, NC.
Blight, Seedling Smother. **Thelephora terrestris**, ID, MN, OH; seedling, **Rosellinia herpotrichioides**, WA.
Blight, Shoot. **Sirococcus strobilinus**, KS, NC, WI; **Phomopsis occulta**, WI.
Blight, Snow. **Phacidium infestans**, New England states; **Botrytis cinerea**, NJ; **Lophophacidium hyperboreum**, MN.
Blight, Twig. **Ascochyta piniperda**, ME, NC.
Brooming. **Rhizosphaera kalkhoffii**, Central Rocky Mts.
Canker, Bark. **Aleurodiscus amorphus**, widespread.
Canker; Twig Blight. **Cytospora kunzei** (*Valsa kunzei* var. *piceae*), New England to NJ, IL, MN.

Damping-Off. **Cylindrocladium scoparium**, MI, NJ; **Phytophthora cinnamomi**, MD, NY, VA; **P. cactorum**, NY, VA; **Pythium ultimum**, widespread; **Rhizoctonia solani**, cosmopolitan; **Aphanomyces euteiches**; **Caloscypha fulgens**, OR, WA.
Dieback, Branch. **Phomopsis occulta**, MI.
Mistletoe, White Fir Dwarf. **Arceuthobium abietinum** f. sp. **concoloris**, MS.
Mistletoe; Witches' Broom. **Arceuthobium campylopodium**, Rocky Mts., NM, TX; **A. pusillum**, New England, especially ME, NH, to Great Lakes states, MN, WI; **A. vaginatum**, CO.
Needle Cast. **Rhizosphaera kalkhoffii**, northeastern U.S., AZ, IN, MI, MN, PA, WI; **Stigmina lautii**, ND.
Needle Cast; Tar Spot. **Lophodermium filiforme**, AK, CO, NY; **L. piceae**, New England to Great Lakes states, Pacific Northwest; **Lophodermina septata**, OR; **Rhizosphaera kalkhoffii**, CT, NY, VA; **Bifusella crepidiformis**, MT; **Cladosporium** sp., ME.
Nematode. **Nacobbodera chitwoodi**, OR; **Paratylenchus projectus**; **Pratylenchus penetrans**; **Tylenchus marginatus**.
Nematode, Dagger. **Xiphinema americanum**, WI.
Rot, Collar. **Diplodia pinea**, NJ, NY; **Sphaeropsis ellisii**, NJ.
Rot, Heart. **Fomes annosus**, CA to WA; **F. pini**, widespread; **F. pinicola**, widespread; **F. roseus**; **F. subroseus**, wide spread; **Polyporus** spp.
Rot, Root. **Armillaria mellea**, AZ, CO, NM, OR, PA, UT, WA; **Inonotus tomentosus**; **Phymatotrichum omnivorum**, TX; **Sparassis radicata**, ID, MT, OR, WA; **Phytophthora cinnamomi**, NC; **P. dreschleri**, NC; **P. cryptogea**, NC; **Fusarium solani**, PA; **F. oxysporum**, PA; **F. avenaceum**, PA; **Cylindrocladium canadense**, Montreal, Canada.
Rot, Root, Butt. **Hyphodontia aspera**, NH.
Rust, Cone. **Chrysomyxa pirolata** (0, I), AK, CO, ME, MA, MI, MT, NH, NY, OR, PA, VT; II, III on *Moneses* and *Pyrola*.
Rust, Needle. **Chrysomyxa empetri** (0, I), ME; **C. ledi** (0, I), New England to Great Lakes; II, III on *Ledum*; **C. ledi** var. **cassandrae** (0, I), New England to Great Lakes; II, III on *Chamaedaphne*; **C. ledicola** (0, I), northern U.S. from ME to WA, AK, CO; II, III on *Ledum*; wild rosemary; **C. piperiana** (0, I), CA, OR; II, III on *Rhododendron californicum*, **C. roanensis** (0, I), NC, TN; II, III on *Rhododendron catawbiense*, **C. weirii**, TN, WV to VT, WI, SD to WA, most Canadian Provinces; **C. arctostaphyli** (0, I); III on bearberry.
Rust; Witches' Broom. **Melampsorella cerastii**, widespread.

Cytospora canker frequently kills lower branches of ornamental spruces. There is little control except to remove affected portions.

SQUASH-BUSH (*Condalia*)
Mistletoe. **Phoradendron californicum**, CA.

SQUAW-APPLE (*Peraphyllum*)
Leaf Spot. **Septoria peraphylli**, UT.
Rust. **Gymnosporangium inconspicuum** (0, I), CO; III on juniper; **G. nelsonii** (0, I), on leaves, fruit; CO, UT.

STAPHYLEA (Bladdernut)
BLIGHT, Twig. **Hypomyces ipomoeae**, MA; **Coryneum microstictum**, MA.
LEAF SPOT. **Mycosphaerella staphylina**, GA, KS. **Ovularia isarioides**, NY to MO, IA; **Septoria cirrhosa**, MO.

STENOLOBIUM (Florida Yellow-Trumpet)
ROT, Root. **Clitocybe tabescens**, FL; **Phymatotrichum omnivorum**, TX.
RUST. **Prospodium appendiculatum** (0, I, II, III), FL, TX.

STRANVAESIA
BACTERIAL Fire Blight. **Erwinia amylovora**, NJ.
ROT, Root. **Clitocybe tabescens**, FL.

STRAWBERRY-TREE (*Arbutus unedo*)
BACTERIAL Crown Gall. **Agrobacterium tumefaciens**, CA.
LEAF SPOT. **Septoria unedonis**, OR.
SPOT ANTHRACNOSE. **Elsinoë mattirolianum**, CA.

STYRAX (Snowbell)
LEAF SPOT. **Mycosphaerella punctiformis**, GA.
NEMATODE, Root Knot. **Meloidogyne** sp., MD.

SUMAC (*Rhus*)
CANKER, Stem; DIEBECK. **Botryosphaeria ribis**, "umbrella disease," NY to GA, MN, NE; **Cryptodiaporthe aculeans**, ME to AL, IA, OK; **Nectria cinnabarina**, NY, OK; **N. galligena**, NY, PA, VA; **Physalospora obtusa**, canker, inflorescence blight, NY to AL, KS, MI.
DODDER. **Cuscuta exaltata**, TX.
LEAF BLISTER. **Taphrina purpurascens**, MA to GA, KS, TX.
LEAF SPOT. **Cercospora rhoina**, general; **Pezizella oenotherae**, NY to GA, WV; **Coniothyrium rhois**, TX; **Harknessia rhoina**, CA; **Phyllosticta rhoiseda**, CA; **Septoria rhoina**, general; **Cylindrosporium** sp., CO, NE.
MOLD, Leaf. **Cladosporium aromaticum**, IA, NE, NY, WI, WV.
PARASITIC LICHEN. **Strigula elegans** and **S. complanata**, LA, Southern U.S.
POWDERY MILDEW. **Sphaerotheca macularis**, widespread; **Oidium** sp., WI.
ROT, Root. **Armillaria mellea**, CA; **Clitocybe tabescens**, FL; **Corticium galactinum**, white root, VA; **Phymatotrichum omnivorum**, TX.
ROT, Wood. **Polyporus** spp.; **Poria punctata**, MI; **Schizophyllum commune**, cosmopolitan; **Steccherinum ochraceum**, PA.
RUST. **Pileolaria effusa** (0, III), AZ; **P. patzcuarensis** (0, I, II, III), CO, NM, OK.
WILT. **Fusarium oxysporum** f. sp. **rhois**, CT, KS, VA; **Verticillium alboatrum**, IL, IA, MA, MN.

SWEET-FERN (*Comptonia*)
RUST, Blister. **Cronartium comptoniae** (II, III), ME to MN, NC, OH; 0, I on hard pine.
RUST, Leaf. **Gymnosporangium ellisii** (0, I), NJ; II, III on *Chamaecyparis*.

SWEET GALE (*Myrica gale*)
BLIGHT, Twig. **Diplodia** sp., NJ.
LEAF SPOT. **Ramularia monilioides**, NY: **Septoria myricata**, NY.
RUST. **Cronartium comptoniae**, ME to NY, WA; 0, I on pine; **Gymnosporangium ellisii** (0, I), ME.

SWEET GUM (*Liquidambar*)
ANTHRACNOSE. **Gloeosporium nervisequum**.
BACTERIAL LEAF SCORCH. **Xylella fastidiosa**, KY.
BLIGHT; Leader Dieback. Cause unknown, killing trees in AL, AR, FL, GA, LA, MD, SC, TX.
BLIGHT, Thread. **Pellicularia koleroga**, LA.
CANKER; DIEBACK. **Botryosphaeria ribis** var. **chromogena**, IL, NY, MD to FL, LA; **B. dothidea**, MS; **Hymenochaete agglutinans**, MD; **Nectria** sp.; **Dothiorella berengeriana**, bleeding necrosis, NJ, NY; **Xylella fastidiosa**, Southern CA.
CANKER, Felt Fungus. **Septobasidium alni**; **S. apiculatum**, **S. burtii**, **S. mariani**, **S. pseudopedicellatum**, and **S. sinuosum**, southern states.
LEAF SPOT. **Cercospora liquidambaris**, MD to FL, TX; **C. tuberculans**, FL, LA, MS, MO; **Discosia artocreas**, OK; **Exosporium liquidambaris**, TX; **Leptothyrella liquidambaris**, red leaf spot, IL, MD, NC; **Septoria liquidambaris**, MA to FL, TX; **Actinopelte dryina**; **Cladosporium** sp., OR.
MISTELTOE. **Phoradendron serotinum** (**flavescens**), OH to NC, TX.
NEMATODE, Root Knot. **Meloidogyne** sp.
NONPARASITIC. **Blight**; **Decline**; Leader Dieback. Killing trees in AL, AR, FL, GA, LA, MD, SC, TX. Partly moisture shortage.
ROT, Heart. **Polyporus adustus** and **P. gilvus**, widespread.
ROT, Root. **Clitocybe tabescens**, FL; **Phymatotrichum omnivorum**, TX; **Cylindrocladium scoparium**, GA.
ROT, Wood. **Daedalea confragosa**, cosmopolitan; **Fomes** spp., **Ganoderma** spp.; **Hericium erinaceus**; **Steccherinum ochraceum**; **S. pulcherrimum**, Gulf states; **Polyporus** spp., **Poria** spp.; **Pleurotus** spp.; **Schizophyllum commune**, cosmopolitan; **Stereum** spp.

SWEETLEAF (*Symplocos*)
GALL, Bud. **Exobasidium symploci**, Gulf states to NC and IN.
LEAF SPOT. **Septoria stigma**, AL; **S. symploci**, FL, MS, NC, OK, TX.
ROT, Root. **Phymatotrichum omnivorum**, TX.

SWEET-OLIVE (*Osmanthus fragrans*)
BLACK MILDEW; Black Spot. **Asterina** sp., MS.
LEAF SPOT. **Gloeosporium oleae**, MD.
ROT, Root. **Armillaria mellea**, CA; **Clitocybe tabescens**, LA.

SYCAMORE
See Plane Tree

T

TABEBUIA (Trumpet-Tree)
RUST. **Prospodium plagiopus** (II, III), FL.

TABERNAEMONTANA (Crape-Jasmine)
LEAF SPOT. **Gloeosporium tabernaemontanae**, FL.
LEAF SPOT, Algal; Green Scurf. **Cephaleuros virescens**, FL.
LEAF MOLD. **Cladosporium** sp., FL.
ROT, Mushroom Root. **Clitocybe tabescens**, FL.
RUST. **Uredo manilensis**, FL.

TAMARIND (*Tamarindus*)
NEMATODE, Root Knot. **Meloidogyne** sp., FL.

TAMARISK, SALT CEDAR (*Tamarix*)
BACTERIAL Blight. Fatal to some plants in AZ, NM, TX, where the host has become a noxious weed.
POWDERY MILDEW. **Sphaerotheca macularis**, IN.
ROT, Root. **Phymatotrichum omnivorum**, CA, TX.
ROT, Wood. **Polyporus sulphureus**, MD.

TEA (Thea sinensis)
BLIGHT, Twig. **Pestalotia guepini**, also leaf spot, SC.
LEAF SPOT, Algal; Green Scurf. **Cephaleuros virescens**, FL, SC.
SPOT ANTHRACNOSE; Scab. **Sphaceloma** sp., LA; **Elsinoë leucospila**, FL.

TERNSTROEMIA
SPOT ANTHRACNOSE. **Elsinoë leucospila**, FL.

THUJOPSIS (Hiba Arborvitae)
BLIGHT, Twig. **Phomopsis occulta**, CA.

THUNBERGIA (Clockvine)
BACTERIAL Crown Gall. **Agrobacterium tumefaciens**, CT, FL.
BACTERIAL, MLO. **Aster Yellows**, NY.

R.K. Horst, *Field Manual of Diseases on Trees and Shrubs*,
DOI 10.1007/978-94-007-5980-0_23, © Springer Science+Business Media Dordrecht 2013

NEMATODE, Root Knot. **Meloidogyne** sp., FL.
VIRUS. **Broad Bean Wilt**, NY.

THYME (*Thymus*)
BLIGHT, Gray Mold. **Botrytis cinerea**, AK.
LEAF SPOT. **Corynespora cassiicola**, FL.
ROT, Root. **Rhizoctonia solani**, MA.

TIBOUCHINA (Glory-Bush)
ROT, Mushroom Root. **Clitocybe tabescens**, FL.

TIDESTROMIA
ROT. **Macrophomina phaseolina**, AZ.

TITHONIA
VIRUS. **Tomato Spotted Wilt-Lettuce Serotype**, GA.

TORREYA
LEAF SPOT. **Phomopsis** sp.; **Diaporthe**, **Lasiodiplodia**, **Hypoxylon**, **Fuarium** spp. **Botryosphaeria obtuca**, FL, GA.
NEEDLE SPOT. **Fusarium lateritium**, FL.
NEEDLE SPOT, CANKER STEM. **Pestalotiopsis microspora**, FL.

TREE-POPPY (*Dendromecon*)
SMUT, Leaf. **Entyloma eschscholtziae**, CA.

TREE-TOMATO (*Cyphomandra*)
BACTERIAL Canker. **Clavibacter michiganense**, CA.
POWDERY MILDEW. **Oidium** sp., MD.

TROPICAL SODA APPLE (*Solanum viarum*)
NEMATODE, Root Knot. **Meloidogyne arenaria**, FL.

TULIP-TREE, YELLOW POPLAR (*Liriodendron*)
BACTERIAL Root Lesion. **Pseudomonas** sp., OR.
BLIGHT, Seedling. **Rhizoctonia solani**, OH, VA.
CANKER. **Dothiorella** sp., PA; **Myxosporium** spp., NY; **Nectria** sp., WV to NC, TN; **Fusarium solani**, SC.
LEAF SPOT. **Cylindrosporium cercosporioides**, MD, WV; **Gloeosporium liriodendri**, CT to NJ, TX; **Mycosphaerella liriodendri**, GA, MI; **M. tulipiferae**, Middle Atlantic and Gulf states; **Phyllosticta lirodendrica** (conidial stage of *Mycosphaerella*), widespread; **Ramularia liriodendri**, AL, DE.
LEAF SPOT, Tar, Black. **Ectostroma liriodendri**, widespread; **Rhytisma liriodendri**, CA, TX, VA.
NEMATODE, Lesion. **Pratylenchus pratensis**, TX.
POWDERY MILDEW. **Phyllactinia corylea**, NY to AL, MO; **Erysiphe polygoni**, widespread.
ROT, Heart. **Collybia velutipes**, WV; **Ganoderma applanatum**, occasional.
ROT, Root. **Armillaria mellea**, VA; **Phymatotrichum omnivorum**, TX; **Cylindrocladium liriodendri**, CA.

Rot, Sapwood; Wood. **Daedalea extensa** and **D. unicolor**, sometimes on standing trees; **Daldinia vernicosa**, cosmopolitan; **Polyporus** spp.; **Schizophyllum commune**, wound rot; also many stain and timber rots.
Rot, Seedling Collar. **Cylindrocladium scoparium**, NJ, NC, TN.
Sooty Mold. **Capnodium elongatum**, cosmopolitan.
Virus. **Chlorotic Ringspot** and **Vein Banding**, NC, SC.

The most conspicuous fungus on ornamental tulip-trees is the black sooty mold growing in copious honeydew secreted by tulip-tree aphids and scales. In hot, dry weather leaves sometimes turn yellow and drop prematurely. Leaf spots are seldom serious enough for treatment.

TUNG TREE (*Aleurites*)
Anthracnose. **Glomerella cingulata**, FL.
Blight, Southern. **Sclerotium rolfsii**, TX.
Blight, Thread. **Pellicularia koleroga**, FL, LA, MS, TX; Web, **P. filamentosa**, LA, MS.
Canker; Dieback; Nut Rot. **Botryosphaeria ribis**, GA, LA; **Physalospora rhodina**, FL, LA, MS, TX.
Canker, Felt Fungus. **Septobasidium pseudopedicellatum**, LA.
Leaf Spot. **Cercospora websteri**, MS; **Gloeosporium aleuriticum**, MS; **Phyllosticta** sp., FL, GA, **Cristulariella pyramidalis**, FL.
Leaf Spot, Angular. **Mycosphaerella** (*Cercospora*) **aleuritidis**, MS.
Nematode, Root Knot. **Meloidogyne** sp., AL, FL, MS.
Nonparasitic. **Chlorosis**. Copper or potassium deficiency, FL.
Frenching. Manganese deficiency, FL.
Bronzing. Zinc deficiency, FL.
Wetwood. Alcoholic slime flux. Perhaps bacterial in part.
White Seed. Genetic abnormality.
Rot, Root, Collar. **Clitocybe tabescens**, serious throughout tung belt; **Cephalosporium** sp., LA; secondary; **Phytophthora cinnamomi**, LA.
Rot, Root; Collar Rot. **Cylindrocladium scoparium**, **C. crotalariae**, **C. floridanum**, MS. Spraying is usually necessary for thread blight and nut rot.

TUPELO, SOUR GUM, BLACK GUM (*Nyssa*)
Anthracnose. **Colletotrichum acutatum**, GA.
Blight, Thread. **Pellicularia koleroga**, LA.
Canker. **Nectria galligena**, CT; **Strumella coryneoidea**, northern Appalachians; **Botryosphaeria ribis**, IL; **Fusarium solani**, LA.
Canker, Felt Fungus. **Septobasidium** spp., NJ to Gulf states.
Leaf Spot. **Actinopelte dryina**, AL, IL, OK; **Cercospora nyssae**, TX; **Mycosphaerella nyssaecola**, MA to GA, MI; **Phyllosticta nyssae**, southeastern states to TX.
Mistletoe. **Phoradendron serotinum (flavescens)**, FL, IN, MD, TX.
Rot, Heart. **Ganoderma applanatum**; **Fomes connatus**, CT; **Hericium erinaceus**, NC.

ROT, Root. **Phymatotrichum omnivorum**, TX.
ROT, Wood. **Daedalea confragosa**; **D. unicolor**, occasional; **Lentinus** spp., cosmopolitan; **Polyporus** spp.; **Stereum** spp.; **Trametes rigida**, Gulf states.
RUST. **Aplopsora nyssae** (II, III), MD to AL, TX.
WILT, Seedling. **Phytophthora cactorum**, MO.

TUPIDANTHUS
LEAF SPOT. **Alternaria panax**, CA.

TURPENTINE TREE (*Syncarpia*)
NEMATODE, Root Knot. **Meloidogyne** sp., FL.

U

UDO (*Aralia cordata*)
BLIGHT. **Alternaria** sp., DE, NJ.
ROT, Stem. **Sclerotinia sclerotiorum**, MD.
WILT. **Verticillium albo-atrum**, ME, PA.

UMBRELLA-PINE (*Sciadopitys*)
BLIGHT, Twig. **Diplodia pinea**, NJ.
LEAF SPOT. **Phyllosticta** sp., RI; **Cytospora** sp., OR.
ROT, Root; Damping-off. **Rhizoctonia solani**, CT.
ROT, Root; Wilt. **Phytophthora cinnamomi**, VA.

R.K. Horst, *Field Manual of Diseases on Trees and Shrubs*,
DOI 10.1007/978-94-007-5980-0_24, © Springer Science+Business Media Dordrecht 2013

V

VANCOUVERIA

LEAF SPOT. **Phragmodothis berberidis**, CA; **Ramularia vancouveriae**, CA, OR.

VIBURNUM

BACTERIAL Crown Gall. **Agrobacterium tumefaciens**, PA, WA.

BACTERIAL Leaf Spot. **Pseudomonas viburni**, IL, IA, NJ.

BLIGHT. **Phytophthora ramorum**, OR, WA.

BLIGHT, Gray Mold; Shoot. **Botrytis cinerea**, MA, WA.

BLIGHT, Sudden Oak Death. **Phytophthora ramorum**, CA.

BLIGHT, Thread. **Pellicularia koleroga**, FL, NC.

CANKER, Stem Girdle. **Hymenochaete agglutinans**, PA.

DODDER. **Cuscuta compacta**, FL.

DOWNY MILDEW. **Plasmopara viburni**, general.

GALL, Stem; Dieback. **Phomopsis** sp., MD, NJ, NY.

LEAF SPOT. **Cercospora opuli**, IA, MS; **C. tinea**, LA; **C. varia**, general; **Helminthosporium beaumontii**, AL, TX; **Hendersonia foliorum** var. **viburni**, FL, TX; **H. tini**, LA; **Leptosphaeria tini**, LA; **Monochaetia desmazierii**, WV; **Phyllosticta lantanoides**, NY; **P. punctata**, FL, IA, WI; **P. tinea**, MD; **Ramularia viburni**, TN, WI.

LEAF SPOT. **Cristulariella pyramidalis**, MD.

LEAF SPOT, Algal. **Cephaleuros virescens**, FL, LA.

MOLD, Leaf. **Cladosporium herbarum**, OH, PA.

NEMATODE, Root Knot. **Meloidogyne** spp., CA, MD, MS.

POWDERY MILDEW. **Microsphaera alni**, general.

ROT, Root. **Clitocybe tabescens**, FL; **Corticium falactinum**, MD; **Phymatotrichum omnivorum**, TX; **Rosellinia necatrix**, CA; **Phytophthora cinnamomi** VA.

RUST. **Aecidium rubromaculans** (0, I), FL; **Coleosporium viburni** (II, III), IL, IA, MD, MI, VA, WI; 0, I unknown; **Puccinia linkii** (III), AK, ID, MI, MT, NH, WA.

R.K. Horst, *Field Manual of Diseases on Trees and Shrubs*,

DOI 10.1007/978-94-007-5980-0_25, © Springer Science+Business Media Dordrecht 2013

Spot Anthracnose. **Sphaceloma viburni**, CA, WA.
Wilt. **Verticillium albo-atrum**, IL, IN, OR, WA.

Viburnum carlesii is extremely sensitive to sulfur and may be injured even by spray drift from other plants.

VITEX (*Chaste-Tree*)
Leaf Spot. **Cercospora viticis**, LA, TX.
Rot, Root. **Phymatotrichum omnivorum**, TX.

W

WALNUT (*Juglans*) (Includes Butternut, Black, English, and Japanese Walnuts)

ANTHRACNOSE; Leaf Spot. **Gnomonia leptostyla** (*Marssonina juglandis*), general.

BACTERIAL Blight. **Xanthomonas juglandis**, NY to GA, TX, and Pacific Coast, especially on English (Persian) walnut.

BACTERIAL Canker. **Erwinia nigrifluens**, CA; **E. rubrifaciens**, phloem canker, CA; **Sirococcus clavigignenti-juglandacearum**, NC.

BACTERIAL Crown Gall. **Agrobacterium tumefaciens**, occasional.

BLIGHT, Leaf. **Cylindrosporium juglandis**, AL and TN to TX, CA.

BLIGHT, Seedling. **Sclerotium rolfsii**, TX.

CANKER. **Sirococcus clavigignenti-juglandacearum to Ophiognomonia clavigignenti-juglandacearum**, Ontario, Canada; **Phomopsis oblonga**, CT.

CANKER; Branch Wilt. **Hendersonula toruloidea**, connected with sunburn.

CANKER; DIEBACK. **Diplodia juglandis**, widespread; **Dothiorella gregaria**, CA; **Exosporina fawcetti**, wilt, CA; **Melanconis juglandis**, widespread, especially in East; **Nectria** spp., widespread; **N. galligena**; **Cytospora** sp., AZ.

CANKER; Felt Fungus. **Septobasidium curtisii**, NC.

CANKER, Stem. **Fusarium solani**, KS; **Phytophthora citricola, NC**.

CANKER DISEASE. **Geosmithia**, AZ, CA, CO, ID, OR, UT, WA.

DIEBACK, Twig and Branch. **Neofusicoccum mediterraneum**, CA.

LEAF SPOT. **Cercospora juglandis**, KS, MA; **Phloeospora multimaculans**, TX; **Marssonina californica**, CA; **Ascochyta juglandis**, ring spot; **Grovesinia pyramidalis**, IL, IA, MN, OH, WV; **Mycosphaerella juglandis**, IL, IN, IA, NC; **Cristulariella pyramidalis**, IL. Also on *Bidens frondosa* (beggar-ticks), *Campsis radicans* (trumpet vine), *Chenopodium ambrosioides* (Mexican tea), *Commelina diffusa* (dayflower), *Cuphea petiolota* (blue waxweed), *Desmodium canescens* (tick clover), *Eupatorium coelestinum* (mist-flower), *E. rugosum* (white snakeroot), *Ipomoea hederscea* (morning-glory), *I. lacunosa* (morning-glory), *Lobelia inflata* (Indian tobacco), *L. siphilitica* (blue cardinal-flower), *Perilla frutescens* (beefsteak plant), *Phytolacca arnericana* (poke), *Platanus occidentalis* (sycamore), *Polygonum pennsylvanicum* (smart weed), *P. scandens* (false buckwheat),

R.K. Horst, *Field Manual of Diseases on Trees and Shrubs*,
DOI 10.1007/978-94-007-5980-0_26, © Springer Science+Business Media Dordrecht 2013

Rumex crispus (yellow dock), *Sida spinosa* (prickly mallow), *Solidago canadensis* (goldenrod), *Vitis palmata* (Catbird grape).

LEAF SPOT; Downy Spot; White Mold. **Microstroma juglandis**, wide-spread; **M. brachysporum**, general; **Alternaria arborescens**, **A. alternata**, **A. tenuissima** (leaf spot and brown/gray apical necrosis), AZ.

MISTLETOE. **Phoradendron serotinum (flavescens)**, IN southward, AZ, CA, NM.

MOLD, Nut. **Alternaria** sp., CA.

NEMATODE, Lesion. **Pratylenchus musicola**, CA; **P. vulnus**.

NEMATODE, Root Knot. **Meloidogyne** spp., TX.

NONPARASITIC. **Black End**, of Nuts. Probably drought injury, CA, OR.

Black Line, Girdle. Graft incompatibility.

Dieback. Boron deficiency, OR.

Erinose. Leaf galls from blister mites.

Leaf Scorch, **Sunscald**. NJ. OR, WA.

Little Leaf. Zinc deficiency, CA. Probably also causes Rosette.

Shrivel, Witches' Broom, Yellows. Cause unknown.

PHYTOPLASMA, **Witches' Broom**, GA, IL, IN, IA, OH.

POWDERY MILDEW. **Phyllactinia corylea**, IN, OH, OR; **Microsphaera alni**, widespread; **Erysiphe polygoni**, CA.

ROT, Collar. **Phytophthora cactorum**, CA; **P. cinnamomi**, MD to AL, LA.

ROT, Heart. **Fomes igniarius** and **F. everhartii**, widespread; **Polyporus sulphureus**, widespread.

ROT, Root. **Armillaria mellea**, cosmopolitan; **Phymatotrichum omnivorum**, TX; **Cylindrocladium** sp., TN.

ROT, Wood. **Fomes conchatus**; **Polyporus** spp.; **Poria** spp.; **Schizophyllum commune**, cosmopolitan; **Daedalea confragosa**; **D. quercina**, widespread.

SCAB. **Cladosporium** sp., MD, MN.

VIRUS. **Brooming Disease**, DC, GA, MD, NY.

Bacterial blight is the most serious disease of walnuts on the Pacific Coast, requiring several sprays. A toxin, juglone, has been considered injurious to many shrubs growing in the vicinity of black walnut roots; the toxin also causes vascular wilt-like symptoms in eggplant and tomato in home gardens.

WATER-ELM (*Planera*)

ROT, Wood. **Daedalea ambigua**, SC; **Ganoderma lucidum**.

WAX-MYRTLE, CANDLEBERRY (*Myrica cerifera*)

BLACK MILDEW. **Irene calostroma**, Gulf states; **Irenina manca**, MS; **Meliola manca**, FL.

BLIGHT, Seedling. **Rhizoctonia solani**, NJ.

DODDER. **Cuscuta compacta**, FL.

LEAF SPOT. **Cercospora dispersa**, NJ; **Colletotrichum acutatum**, FL; **Phyllosticta myricae**, NJ to FL, TX; **Septoria myricae**, NJ.

NEMATODE, Ring. **Hemicriconemoides wessoni**, FL.

PARASITIC LICHEN. **Strigula elegans** and **S. complanata**, LA, Southern US.

Rot, Root. **Clitocybe tabescens**, FL; **Phymatotrichum omnivorum**, TX.
Rust. **Gymnosporangium ellisii** (0, I), MA to MD; III on *Chamaecyparis*.
Sooty Mold. **Capnodium grandisporum**, FL.

WEIGELA
Bacterial Crown Gall. **Agrobacterium tumefaciens**, MD, MS.
Blight, Twig. **Phoma weigelae**, WA.
Leaf Spot. **Cercospora weigelae**, MD, MS, NJ; **Ramularia diervillae**, TN.
Nematode, Lesion. **Pratylenchus pratensis.**
Nematode, Root Knot. **Meloidogyne** sp., CA, MD, MS, TX.
Rot, Root. **Phymatotrichum omnivorum**, TX.

WHIPPLEA
Downy Mildew. **Peronospora whippleae**, CA.

WHORTLEBERRY, BILBERRY (*Vaccinium* spp.)
Blight, Twig; Berry Rot. **Monilinia ledi**, NY.
Gall, Leaf. **Exobasidium vaccinii**, occasional; **E. parvifolii**, stem gall, OR, WA; **E. vaccinii-uliginosi**, shoot gall, rose-bloom, OR.
Leaf Spot, Tar. **Rhytisma vaccinii**, AK.
Powdery Mildew. **Microsphaera alni** var. **vaccinii**, AK, OR, WA, WY; **Podosphaera oxyacanthae**, AK.
Root, Rot. **Phytophthora cinnamomi**, VA; **P. nicotianae**, VA; **P. palmivora**, VA.
Rust. **Pucciniastrum** sp. (II), OR, WA; **P. goeppertianum** (III), Witches' Broom, general; 0, I on fir; **P. vaccinii** (II, III), general; 0, I on hemlock.

WILLOW (*Salix*) (Includes Weeping Willow, Pussy Willow)
Bacterial Crown Gall. **Agrobacterium tumefaciens**, CT, NJ, TX, VA.
Bacterial Wetwood. **Erwinia nimipressuralis.**
Blight, Twig. **Diplodia salicina.**
Blight, Willow. Complex of scab and black canker.
Canker, Black. **Physalospora miyabeana**, New England, NY, WV.
Canker, Twig and Branch. **Botryosphaeria ribis**, VA to GA, AR, CA; **Cryptodiaporthe salicina**, ME to VA, AK, CA, KS, OK, SD, WA; **Cryptomyces maximus**, bark blister; NM, UT; **Discella carbonacea**, twig blight, New England; **Dothiora polyspora**, CO; **Dothiorella** sp., AR, ND; **Macrophoma** spp., AR, KY, MS, NC, SC, TX; **Phomopsis salicina**, IA, MA, VA; **Physalospora gregaria**, WV; **Valsa** spp., twig canker; **V. sordida** (*Cytospora chrysosperma*), widespread; **V. salicina**, widespread; **V. nivea**, western states.
Dodder. **Cuscuta** spp., IA, NY, WA.
Leaf Blister. **Taphrina populi-salicis**, CA.
Leaf Spot. **Ascochyta salicis**, CA; **Cercospora salicina**, IL, LA, MD, TX; **Ciborinia foliicola**; **C. wisconsinensis**; **Cylindrosporium salicinum**, MA to CO, MS, WI; **Gloeosporium** spp., also twig blight, CT, DE, MA; **G. salicis**, VT to NJ, MS, OR, WI; **Marssonina** spp., widespread; **M. kriegeriana**, WI; **M. apicalis**, CA; **Myrioconium comitatum**, WI; **Phyllosticta apicalis**, KS, WI; **Ramularia rosea**, CO, MT, WI; **Septogloeum salicinum**, AK, NY, WI; **S. maculans**, CA,

MT; **S. salicis-fendlerianae**, ID; **Septoria** spp.; **S. salicicola**, AK, OR; **S. didyma**, WI.
LEAF SPOT, Tar. **Rhytisma salicinum**, MI.
MISTLETOE. **Phoradendron serotinum (flavescens)**, AZ, CA, IN, NM, TX.
MISTLETOE, European. **Viscum album**, CA.
NEMATODE, Lesion. **Pratylenchus vulnus**, CA.
NEMATODE, Root Knot. **Meloidogyne** sp.
POWDERY MILDEW. **Uncinula salicis**, general; **Phyllactinia corylea**, WA.
ROT, Heart. **Daedalea confragosa**, widespread; **Fomes** spp.; **Polyporus farlowii**, TX to CA; **Trametes suaveolens**, New England to MT, AK.
ROT, Root. **Armillaria mellea**, CA, WA; **Helicobasidium purpureum**, TX; **Phymatotrichum omnivorum**, TX.
ROT, Sapwood; Wood. **Daedalea ambigua**; **Ganoderma lucidum**; **Pholiota** spp.; **Pleurotus** spp.; **Polyporus** spp.; **Schizophyllum commune**, cosmopolitan.
RUST. **Melampsora ableti-capraearum** (*M. epitea*) (II, III), general except far North; 0, I on fir; **M. arctica** (II, III), AK, CO, NH; **M. epitea**, MN; **M. paradoxa** (*M. bigelowii*) (II, III), ME to NC, AK, IA, NM; 0, I on larch; **M. ribesii-purpureae** (II, III), MT to CO, AK, CA; 0, I on *Ribes*.
SCAB, Gray; Blight. **Fusicladium saliciperdum** (*Venturia chlorospora*), New England to NJ, NC, PA.
SOOTY MOLD. **Capnodium** sp.
SPOT ANTHRACNOSE; Scab. **Sphaceloma murrayae**, CA, OR, WA, WI.

Willow scab, followed by black canker, forms a very destructive blight, killing many trees in New England.

WINTERBERRY (*Ilex verticillata*)
LEAF SPOT. **Physalospora ilicis**, NY.
LEAF SPOT, Tar. **Rhytisma concavum**, WI; **R. prini**, ME to MS, IL, WI.
POWDERY MILDEW. **Microsphaera alni**, WI.

WINTERGREEN, CHECKERBERRY (*Gaultheria procumbens*)
ANTRACNOSE. **Colletotrichum gloeosporioides**, British Columbia, Canada.
BLOTCH, Sooty. **Gloeodes pomigena**, WI.
FRUIT SPOT, Black Speck. **Leptothyrium pomi**, WI.
LEAF SPOT. **Cercospora gaultheriae**, NY, WI; **Pezizella oenotherae**, VA; **Discosia maculicola**, secondary; **Mycosphaerella gaultheriae**, ME to MD, WV; **Phyllosticta gaultheriae**, general; **Schizothyrium gaultheriae**, ME to VA, WI; **Venturia arctostaphyli**, MD, MA, NJ, NY, VA.
POWDERY MILDEW. **Microsphaera alni**, WI.

WINTERGREEN, WAXFLOWER
See Pipsissewa.

WITCH-HAZEL (*Hamamelis*)
BACTERIAL Crown Gall. **Agrobacterium tumefaciens**, MD.
LEAF SPOT. **Discosia artocreas**, OK; **Gonatobotryum maculicola**, NH to WV, FL, WI; **Graphium hamamelidis**, NY to IN, TN; **Monochaetia desmazierii**, VA to

GA, TN; **Mycosphaerella** sp., WV; **Phyllosticta hamamelidis**, CT to MS, TN, WI; **Ramularia hamamelidis**, NY to WV, OK, WI.
Powdery Mildew. **Phyllactinia corylea**, MI, WI; **Podosphaera biuncinata**, New England to IL and southward.
Rot, Wood. **Fomes scutellatus**, widespread; **Polyporus** spp.

WOLFBERRY (*Symphoricarpos occidentalis*)
Black Knot. **Dibotryon symphoricarpi**, twig canker, ND.
Leaf Spot. **Cercospora symphoricarpi**, MT, ND, WA; **Septoria symphoricarpi**, IA, MT, ND, WA.
Powdery Mildew. **Microsphaera diffusa**, general.
Rot, Collar. **Fomes ribis**, KS, ND.
Rust. **Puccinia crandallii** (0, I), CO, MT, ND, WY; II, III on grasses; **P. symphoricarpi**, MT.

WOOD ROSE (*Rosa gymnocarpa, R. californica*)
Leaf Spot, Lesions. **Phytophthora gymnocarpa**, CA.

Y

YAUPON (*Ilex vomitoria*)
LEAF SPOT; Tar Spot. **Rhytisma ilicincola**, VA.
ROT, Root. **Phymatotrichum omnivorum**, TX.
SOOTY MOLD. **Capnodium** spp., Gulf states.

YELLOWWOOD (*Cladrastris*)
POWDERY MILDEW. **Phyllactinia corylea**, OK.
ROT, Wood. **Polyporus spraguei**, MD.
VIRUS. **Bean Yellow Mosaic**, NY.
WILT. **Verticillium albo-atrum**, IL.

YERBA SANTA (*Eriodictyon*)
BLOTCH, Sooty. **Coniothecium eriodictyonis**, CA.

YEW (*Taxus*)
BLIGHT, Needle. **Herpotrichia nigra**, ID; **Sphaerulina taxi**, CA, ID, OR, WA; **S. taxicola**, also twig blight, NY, PA; **Alternaria** sp., CT, NY, RI; **Phyllosticta taxi**, VA; **Macrophoma taxi**; **Mycosphaerella taxi**; **Neopeckia coulteri**; **Cryptocline taxicola**, eastern United States.
BLIGHT, Ramorum. **Phytophthora ramorum**, PA.
BLIGHT, Seedling. **Phytophthora cinnamomi**, IN, MD, VA, Pacific Northwest.
BLIGHT, Twig. **Phyllostictina hysterella** (*Physalospora gregaria*); **Pestuloria** sp., PA; **P. funerea**, MA; **Sphaeropsis** sp. NJ; **Botryosphaeria ribis**.
DAMPING-OFF. **Rhizoctonia solani**, CT.
ROT, Heart. **Fomes hartigii** OR; **F. roseus**, ID.
ROT, Root. **Armillaria mellea**, ID; **Phytophthora cinnamomi**, OH; **P. lateralis**, CA.
ROT, Wood. **Polyporus schweinitzii**, ID.

YUCCA (Adams-Needle, Joshua-Tree, Spanish Bayonet)
BLIGHT, Flower. **Cercospora floricola** (*C concentrica*), TX.
BLIGHT, Leaf. **Kellermannia anomala**, secondary, general.
LEAF SPOT, **Cercospora concentrica**, CT, GA, IA, NJ, OK, TX; **Coniothyrium concentricum**, general; **Cylindrosporium angustifolium**, KS, MS, OK, TX; **Diplodia**

R.K. Horst, *Field Manual of Diseases on Trees and Shrubs*,
DOI 10.1007/978-94-007-5980-0_27,

circinans, TX; **Epicoccum asterinum**, TX; **Gloeosporium yuccogenum**, MO, TX; **Neottiospora yuccifolia**, GA, WA; **Pestalozziella yuccae**, secondary, TX; **Phyllosticta** sp., TX; **Stagonospora gigantea**, CA; **Leptosphaeria obtusipora**.

MOLD, Leaf. **Torula maculans**, AZ, CA, SC, TX; **Alternaria tenuis**, leaf rot.

NEMATODE. **Meloidogyne** sp., OR.

PHYTOPLASMA. **Yucca Phyllody Bud Proliferation; Aster Yellows**, TX.

ROT, Stem. **Sclerotium rolfsii**, MD.

RUST. **Puccinia amphigena** (0, I), NE.

Many other fungi are found on dead leaves and stems.

Z

ZAMIA (Coontie)
ALGAE. **Anabaena cycadeae**, in coralloid roots, FL.
NEMATODE. **Meloidogyne** sp., FL.

ZEBRA PLANT (*Aphelandra*)
NONPARASTITIC. **Leaf Crinkle**; **Shortened Internodes**, **Axillary Bud Proliferation**, TX.
ROT, STEM. **Phytophthora parasitica**, FL.

R.K. Horst, *Field Manual of Diseases on Trees and Shrubs*,
DOI 10.1007/978-94-007-5980-0_28, © Springer Science+Business Media Dordrecht 2013

Part IV

List of Land-Grant Institutions and Agricultural Experiment Stations in the United States

For help in diagnosing and controlling plant diseases contact your County Ag Comissioners (CA) or county extension agent, your state Diagnostic Lab, your state Department of Agriculture or your Cooperative Extension Specialist.

Alabama: Auburn University, Auburn, AL 36849. http://www.aces.edu/dept/plantdiagnosticlab/

Alabama Cooperative Extension Service http://www.aces.edu/plantlabbham/

Alaska: University of Alaska and Alaska Department of Agriculture http://www.dnr.state.ak.us/ag/index.htm

Arizona: University of Arizona, Tucson, AZ 85721. http://ag.arizona.edu/PLP/plpext/

Arkansas: University of Arkansas, Fayetteville, AR 72701. http://www.aragriculture.org/pestmanagement/diseases/clinic/default.asp

California: Department of Food and Agriculture, Plant Pest Diagnostic Center. (sgaimari@cdfa.ca.gov)

Colorado: Colorado State University, Fort Collins, CO 80523. http://www.coopext.colostate.edu/jeffco/hort/clinic.htm

Connecticut: University of Connecticut, Storrs, CT 06269. (http://pronewengland.org)

Delaware: University of Delaware, Newark, DE 19716. http://ag.udel.edu/extension/pdc/index.htm

District of Columbia: See Rutgers. http://www.rcs.rutgers.eon/services/

Florida: University of Florida, Gainesville, FL 32611. http://plantpath.ifas.ufl.edu/

Georgia: University of Georgia, Athens, GA 30602. http://www.plant.uga.edu/Extension/Clinics/PDC.htm

Hawaii: University of Hawaii, Honolulu, HI 96822. http://www2.ctahr.hawaii.edu/adsc/

Idaho: University of Idaho, Extension Service, Parma, ID 83660. http://www.uidaho.edu/ag/plantdisease/

Illinois: University of Illinois, Urbana, IL 61802. http://plantclinic.cropsci.uiuc.edu/

Indiana: Purdue University, West Lafayette, IN 47907. http://www.ppdl.purdue.edu

Iowa: Iowa State University, Ames, IA 50011. http://www.isuplantdiseaseclinic.org

R.K. Horst, *Field Manual of Diseases on Trees and Shrubs*,
DOI 10.1007/978-94-007-5980-0_29, © Springer Science+Business Media Dordrecht 2013

Kansas: Kansas State University, Manhattan, KS 66506. http://www.plantpath.k-state.edu/DesktopDefault.aspx?tabid=49

Kentucky: University of Kentucky, Lexington, KY 40546. http://www.ca.uky.edu/agcollee/plantpathology/extension/pdd_lab.html

Louisiana: Louisiana State University, Baton Rouge, LA 70803. http://www.lsuagcenter.com/en/our_offices/departments/PlantPathology_Crops_Physiology/Plant_Disease_Clinic/

Maine: University of Maine, Orono, ME 04473. http://pmo.umext.main.edu

Maryland: University of Maryland, College Park, MD 20742. http://www.plant-clinic.umd.edu/

Massachusetts: University of Massachusetts, Amherst, MA 01003. (umassextension.org)

Michigan: Michigan State University, East Lansing, MI 48824. http://www.pestid.msu.edu/

Minnesota: University of Minnesota, St. Paul, MN 55108. http://www.extension.umn.edu/distribution/cropsystems/DC3170.html

Mississippi: Mississippi Cooperative Extension Service, Mississippi State University, MS 39762. http://www.extensionplantclinics.msstate.edu/

Missouri: University of Missouri, Columbia, MO 65211. http://soilplantlab.missouri.edu/

Montana: Montana State University, Bozeman, MT 59717. http://scarab.msu.montana.edu/ipm/

Nebraska: University of Nebraska, Lincoln, NE 68583. http://plantpath.unl.edu/ppathdiagnostic.htm

Nevada: University of Nevada, or Nevada Department of Agriculture. http://www.nevada.gov/

New Hampshire: University of New Hampshire, Durham, NH 03824. http://ceinfo.unh.edu/Agric/AGPDTS/PlantH.htm

New Jersey: Rutgers, The State University, Milltown, NJ 08850. http://www.rce.rutgers.edu.services/

New Mexico: New Mexico State University, Las Cruces, NM 88003. http://plantclinic.nmsu.edu

New York: Cornell University, Ithaca, NY 14853. http://plantclinic.cornell.edu

North Carolina: North Carolina State University, Raleigh, NC 27695. http://www.ncsu.edu/pdic

North Dakota: North Dakota State University, Fargo, ND 58105. http://www.ag.ndsu.nodak.edu/diaglab/

Ohio: Ohio State University, Columbus, OH 43210. http://ppdc.osu.edu

Oklahoma: Oklahoma State University, Stillwater, OK 74078. http://entoplp.okstate.edu/Pddl/

Oregon: Oregon State University, Corvallis, OR 97331. http://web.science.oregonstate.edu/bpp/Plant_Clinic/index.htm

Pennsylvania: The Pennsylvania State University, University Park, PA 16802. http://www.ppath.cas.psu.edu/extension.html

Puerto Rico: Department of Agriculture, San Juan, PR. http://www.agricultura.gobierno.pr

Rhode Island: University of Rhode Island, Kingston, RI 02881. http://www.uri.edu/ce/ceec/plantclinic.html

South Carolina: Clemson University, Clemson, SC 29634. http://www.clemson.edu/agsrvlb

South Dakota: South Dakota State University, Brookings, SD 57007. http://plantsci.sdstate.edu/planthealth/subpage.cfm?ID=12

Tennessee: University of Tennessee, Nashville, TN 37211. http://soilplantandpest.utk.edu/

Texas: Texas A & M University, College Station, TX 77845. http://plantdiseaselab.tamu.edu/

Utah: Utah State University, Logan, UT 84322. http://utahpests.usu.edu

Vermont: University of Vermont, Burlington, VT 05405. http://pss.uvm.edu/pd/pdc/

Virgin Islands: Virgin Islands Extension Service, Kingshill, St. Croix, VI 00850. http://rps.uvi.edu/CES/index.html

Virginia: Virginia Polytechnic Institute, Blacksburg, VA 24061. http://www.ppws.vt.edu/~clinic/

Washington: WSU Research and Extension Center, Puyallup, WA 98371. http://www.puyallup.wsu.edu/plantclinic/index.html

West Virginia: West Virginia University Extension Service, Morgantown, WV 26506. http://www.mastergardener2008.com/

Wisconsin: University of Wisconsin, Madison, WI 53706. http://pddc.wisc.edu/, http://www.entomology.wisc.edu/entodiag.html

Wyoming: University of Wyoming Cooperative Extension Service, Laramie, WY 82071. http://ces.uwyo.edu/

Ontario: Pest Diagnostic Clinic, University of Guelph, Ontario, Canada, N1H 8J7. http://www.labservices.uoguelph.ca/units/pdc/

Glossary

Acervulus, pl. *Acervuli* A "little heap," an erumpent, cushionlike mass of hyphae bearing conidiophores and conidia, sometimes with setae; characteristic of the Melanconiales.

Acicular Needlelike.

Aeciospore Rust spore formed in an aecium.

Aecium, pl. *Aecia* A cluster-cup, or cuplike fruiting sorus in the rusts.

Aerobic Living or active only in the presence of oxygen.

Amoeboid Not having a cell wall and changing in form like an amoeba.

Allantoid Sausage-shaped.

Alternate Host One or other of the two unlike hosts of a heteroecious rust.

Anamorph State of life-cycle in which asexual spores, or none, are produced.

Annulus A ring; ringlike partial veil around stipe in the mushrooms.

Antheridium, pl. *Antheridia* Male sex organ in the fungi.

Anthracnose A disease with limited necrotic lesions, caused by a fungus producing nonsexual spores in acervuli.

Antibiotic Damaging to life; especially a substance produced by one microorganism to destroy others.

Apothecium, pl. *Apothecia* The cup-or saucer-like ascus-bearing fruiting body; in the Discomycete section of the Ascomycetes.

Appressorium A swelling on a fungus germ tube for attachment to host in early stage of infection; found especially in anthracnose fungi and rusts.

Ascocarp, or *Ascoma* Any structure producing asci, as an apothecium, perithecium.

Ascomycetes One of the three main groups of the fungi, bearing sexual spores in asci.

Ascospore Produced in ascus by free cell formation.

Ascus, pl. *Asci* Saclike, usually clavate cell containing ascospores, typically eight.

Aseptate Without cross-walls.

Asexual Vegetative, having no sex organs or sex spores; the imperfect stage of a fungus.

Autoecious Completing life cycle on one host; term used in rusts.

R.K. Horst, *Field Manual of Diseases on Trees and Shrubs*,
DOI 10.1007/978-94-007-5980-0, © Springer Science+Business Media Dordrecht 2013

Bacteria Microscopic one-celled organisms increasing by fission.
Bactericide Substance causing death of bacteria.
Basidiomycetes Class 3 in the Fungi, characterized by septate mycelium, sometimes with clamp-connections, and sexual spores on basidia.
Basidiospore Spore produced on a basidium.
Basidium, pl. *Basidia* Club-shaped structure, which, after fusion of two nuclei, produces four basidiospores.
Binucleate Having two nuclei.
Blight A disease with sudden, severe leaf damage and often with general killing of flowers and stems.
Blotch A blot or spot, usually superficial.
Breaking, of a virus Loss of flower color in a variegated pattern, especially in tulips.

Canker A lesion on a stem; a plant disease with sharply limited necrosis of the cortical tissue.
Carrier Infected plant showing no marked symptoms but source of infection for other plants.
Catenulate In chains, or in an end-to-end series.
Cerebroid With brainlike convolutions or folds.
Chemotherapy Treatment of internal disease by chemical agents that have a toxic effect on the microorganism without injuring the plant.
Chlamydospore Thick-walled, asexual resting spore formed by the rounding up of any mycelial cell; also used for smut spores.
Chlorosis Yellowing of normally green tissue due to partial failure of chlorophyll to develop; often due to unavailability of iron.
Cilium, pl. *Cilia* Hairlike swimming organ on bacteria or zoospores.
Cirrhus, pl. *Cirrhi* A tendril or horn of forced-out spores.
Clamp-connections Outgrowths of hyphae which form bridges around septa, thus connecting two cells; in Basidiomycetes.
Clavate Club-shaped.
Cleistothecium, pl. *Cleistothecia* A perithecium without a special opening; in powdery mildews.
Coalesce Growing together into one body or spot.
Coenocytic Multinucleate; mycelia having no cell walls.
Columella Sterile central axis in a mature fruiting body.
Concentric One circle within another with a common center.
Conidiophore Simple or branched hyphae on which conidia are produced.
Conidium, pl. *Conidia* Any asexual spore except sporangiospore or chlamydospore.
Conk Term used in forestry for sporophores of Polyporaceae on trees.
Control Prevention of, or reduction of loss from, plant disease.
Coremium, pl. *Coremia* Synnema, a cluster of erect hyphae bearing conidia.
Coriaceous Like leather in texture.
Culturing Artificial propagation of organisms on nutrient media or living plants.
Cystidium, pl. Cystidia Sterile, often swollen cell projecting from hymenium in Basidiomycetes.

Damping-off Seed decay in soil, or seedling blight.

Decumbent Resting on substratum with ends turned up.

Decurrent Running down the stipe or stem.

Diagnosis Identification of nature and cause of a disease.

Dieback Progressive death of branches or shoots beginning at tips.

Defoliate To strip or become stripped of leaves.

Dichotomous Branching, frequently successive, into two more or less equal arms.

Dimidiate Having one half smaller than the other; of a perithecium, having outer wall covering only top half.

Discomycetes The cup fungi, a subclass of Ascomycetes; with apothecia.

Disease A condition in which use or structure of any part of the living organism is not normal.

Disinfection Freeing a diseased plant, organ, or tissue from infection.

Disinfestation Killing or inactivating disease organisms before they can cause infection; on surface of seed or plant part, or in soil.

Dissemination Transport of inoculum from a diseased to a healthy plant.

Disjunctor Cell or projection connecting spores of a chain.

Duster Apparatus for applying fungicides in dry form.

Echinulate Having small, pointed spines; used of spores.

Endoconidium, pl. *Endoconidia* Conidium formed within a hypha.

Enphytotic A plant disease causing constant damage from year to year.

Epiphytotic Sudden and destructive development of a plant disease over an extensive area, an epidemic.

Eradicant Fungicide One that destroys a fungus at its source.

Eradication Control of disease by eliminating the pathogen after it is already established.

Erumpent Breaking through surface of substratum.

Excentric Off center.

Exclusion Control of disease by preventing its introduction into disease-free areas.

Exudate Liquid discharge from diseased tissues.

Fasciation Joining side by side; a plant disease with flattened and sometimes curved shoots.

Fascicle A small bundle or cluster.

Filiform Threadlike.

Fimbriate Fringed, or toothed.

Flag A branch with dead leaves on an otherwise green tree.

Flagellum, pl. *Flagella* Whiplike organ on a motile cell; cilium.

Fruiting Body Fungus structure containing or bearing spores; mushroom, pycnidium, perithecium, apothecium, etc.

Fumigant A volatile disinfectant, destroying organisms by vapor.

Fungicide Chemical or physical agent that kills or inhibits fungi.

Fungi Imperfecti Fungi that have not been connected with the perfect or sexual stage; most are imperfect states of Ascomycetes.

Fungistatic An agent preventing development of fungi without killing them.

Fungus, pl. *Fungi* An organism with no chlorophyll, reproducing by sexual or asexual spores, usually with mycelium with well-marked nuclei.

Fusiform Spindle-like, narrowing toward the ends.

Fusoid Somewhat fusiform.

Gall Outgrowth or swelling, often more or less spherical,of unorganized plant cells as result of attack by bacteria, fungi, or other organisms.

Gametangium Gamete mother cell.

Gamete A sex cell, especially one formed in a gametangium.

Germ Tube Hypha produced by a germinated fungus spore.

Gill Lamella or hymenium-covered plate on underside of cap of a mushroom.

Girdle A canker that surrounds stem, completely cutting off water supply and thus causing death; girdling roots also cause death.

*Glabrou*s Smooth.

Gleba Sporulating tissue in an angiocarpous fruit body.

Globose Almost spherical.

Gram-negative, Gram-positive Not being stained, and being stained, by the Gram stain used in classifying bacteria.

Haustorium, pl. *Haustoria* Special hyphal branch extended into living cell for purpose of absorbing food.

Heteroecious Undergoing different parasitic stages on two unlike hosts, as in the rusts.

Heterothallic Of a fungus, sexes separate in different mycelia.

Holocarpic Having all the thallus used for a fruiting body.

Homothallic Both sexes present in same mycelium.

Host Any plant attacked by a parasite.

Hyaline Colorless, or nearly transparent.

Hymenium Spore-bearing layer of a fungus fruiting body.

Hyperplastic Term applied to a disease producing an abnormally large number of cells.

Hypha, pl. *Hyphae* Single thread of a fungus mycelium.

Hypoplastic Term applied to a disease with subnormal cell production.

Hyphopodium, pl. *Hyphopodia* More or less lobed appendage to a hypha.

Hysterothecium Oblong or linear perithecium, sometimes considered an apothecium, opening by a cleft.

Immune Exempt from disease; having qualities that do not permit infection.

Immunization Process of increasing the resistance of a living organism.

Imperfect Fungus One lacking any sexual reproductive state.

Imperfect State State of life-cycle in which asexual spores, or none, are produced. See Anamorph.

Incubation Period Time between inoculation and development of symptoms that can be seen.

Indehiscent Of fruit bodies, not opening, or with no special method.

Infection Process of beginning or producing disease.

Infection Court Place where an infection may take place, as leaf, fruit, petal, etc.

Injury Result of transient operation of an adverse factor, as an insect bite, or action of a chemical.

Innate Bedded in, immersed.

Inoculation Placing of inoculum in infection court.

Inoculum Pathogen or its part, as spores, fragments of mycelium, etc., that can infect plants.

Inoperculate Not opening by a lid.

Intercellular Between cells.

Intracellular Within cells.

Intumescence Knoblike or pustulelike outgrowth of elongated cells on leaves, stems, etc., caused by environmental disturbances.

Lamella Gill.

Lesion Localized spot of diseased tissue.

Locule A cavity, especially one in a stroma.

Macroconidia Large conidia.

Macroscopic Large enough to be seen with the naked eye.

Medulla Loose layer of hyphae inside a thallus; body of a sclerotium.

Microconidia Very small spores, now considered spermatia of a fungus also having larger conidia.

Micron 1/1,000 mm, unit used for measuring spores.

Microscopic Too small to be seen except with the aid of a microscope; true of most of the fungus structures shown in line drawings in this book.

Mildew Plant disease in which the pathogen is a growth on the surface.

Molds Fungi with conspicuous mycelium or spore masses, often saprophytes.

Monoecious Male and female reproductive organs in same individual; in rusts, all stages of life cycle on single species of plant.

Multinucleate Several nuclei in same cell.

Mummy Dried, shriveled fruit, result of disease.

Muriform Having cross and longitudinal septa.

Mushroom An agaric fruit body.

Mycelium, pl. *Mycelia* Mass of fungus hyphae.

Mycelia Sterilia Fungi Imperfecti where spores, except for chlamydospores, are not present.

Mycoplasmalike Organism (MLO) A walless prokaryotic plant pathogen which has a single-unit membrane. See Phytoplasma.

Mycorrhiza, pl. *Mycorrhizae* Symbiotic, nonpathogenic association of fungi and roots.

Necrosis Death of plant cells, usually resulting in tissue turning dark.

Necrotic As an adjective, killing.

Nematicide Chemical or physical agent killing nematodes.

Nematodes Nemas, roundworms, eelworms, cause of some plant diseases.

Obligate Parasite A parasite that can develop only in living tissues, with no saprophytic stage.

Obovate Inversely ovate, narrowest at base.

Obtuse Rounded or blunted, greater than a right angle.

Oogonium, pl. *Oogonia* Female sex organ in the Oomycetes.

Oomycetes Subclass of the Phycomycetes, gametangia of unequal size.

Oospore Resting spore formed in a fertilized oogonium.

Operculate With a cover or lid, as in some asci.

Ostiole Porelike mouth or openings in papilla or neck of a perithecium or pycnidium.

Papilla, pl. *Papillae* Small, nipplelike projection.

Paraphysis, pl. *Paraphyses* A sterile hyphal element in the hymenium, especially in the Ascomycetes, usually clavate or filiform.

Paraphysoids Threads of hyphal tissue between asci, like delicate paraphyses but without free ends.

Parasite An organism that lives on or in a second organism, usually causing disease in the latter.

Pathogen Any organism or factor causing disease.

Pathogenic Capable of causing disease.

Pedicel Small stalk.

Perfect State Stage of life-cycle in which spores are formed after nuclear fission. See Teleomorph.

Peridium Wall or limiting membrane of a sporangium or other fruit body, or of a rust sorus.

Perithecium Subglobose or flasklike ascocarp of the Pyrenomycetes.

Phialide A cell that develops one or more open ends from which a basipetal succession of conidia develops without an increase in length of the phialide itself.

Physiogenic Disease Caused by unfavorable environmental factors.

Physiologic Races Pathogens of same variety and species structurally the same but differing in physiological behavior, especially in ability to parasitize a given host.

Phytopathology Plant pathology, science of plant disease.

Phytoplasma A MLO that is pathogenic to plants.

Pileus Hymenium-supporting part of a fruit body of a higher fungus; the cap of a mushroom.

Primary Infection First infection by a pathogen after going through a resting or dormant period.

Prokaryotic Organisms which lack a true nucleus; includes bacteria and mycoplasmalike organisms.

Promycelium Basidium of rusts and smuts.

Pulvinate Cushionlike in form.

Pycnidium, pl. *Pycnidia* Flasklike fruiting body containing conidia.

Pycnium Spermagonium in the rusts, the 0 stage, resembling a pycnidium.

Resistance Ability of a host plant to suppress or retard activity of a pathogen.

Resting Spore A spore, often thick-walled, that can remain alive in a dormant condition for some time, later germinating and capable of initiating infection.

Resupinate Flat on the substratum with hymenium on outer side.

Rhizoid Rootlike structure.

Rhizomorph A cordlike strand of fungus hyphae.

Ring Spot Disease symptoms characterized by yellowish or necrotic rings with green tissue inside the ring, as in virus diseases.

Roguing Removal of undesired individual plants.

Rosette Disease symptom with stems shortened to produce a bunchy growth habit.

Russet Brownish roughened areas on skins of fruit, from abnormal production of cork caused by disease, insect, or spray injury.

Rust A fungus, one of the Uredinales, causing a disease also known as rust.

Saprophyte An organism that feeds on lifeless organic matter.

Scab Crustlike disease lesion; or a disease in which scabs are prominent symptoms.

Sclerotium, pl. *Sclerotia* Resting mass of fungus tissue, often more or less spherical, normally having no spores in or on it.

Scorch Burning of tissue, from infection or weather conditions.

Scutellum Plate or shieldlike cover, as in Microthyriales.

Septate Having cross-walls, septa.

Sessile Having no stem.

Seta, pl. *Setae* A stiff hair, or bristle, generally dark-colored.

Shot Hole A disease symptom in which small round fragments drop out of leaves, making them look as if riddled by shot.

Sign Any indication of disease other than reaction of the host plant – pores, mycelium, exudate, or fruiting bodies of the pathogen.

Slurry Thick suspension of chemical; used for seed treatment.

Smut A fungus of the Ustilaginales, characterized by sooty spore masses; the name also used for the disease caused by the smut.

Sooty Mold Dark fungus growing in insect honeydew.

Sorus, pl. *Sori* Fungus spore mass, especially of rusts and smuts; occasionally, a group of fruiting bodies.

Species One sort of plant or animal; abbreviated as "sp." singular, and "spp." plural. A genus name followed by sp. means that the particular species is undetermined. Spp. following a genus name means that several species are grouped together without being named individually.

Spermagonium Walled structure in which spermatia are produced, a pycnium.

Spermatium, pl. *Spermatia* A sex cell (+ or –), a pycniospore.

Sporangiole Small sporangium without a columella and with a small number of spores.

Sporangiophore Hypha bearing a sporangium.

Sporangium Organ producing nonsexual spores in a more or less spherical wall.

Spore A single-to many-celled reproductive body, in the fungi and lower plants, which can develop a new plant.

Sporidium, pl. *Sporidia* Basidiospore of rusts and smuts.

Sporodochium, pl. *Sporodochia* Cluster of conidiophores interwoven on a stroma or mass of hyphae.

Sporophore Spore-producing or supporting structure – fruit body; used especially in the Basidiomycetes.

Sporulate To produce spores.

Sprayer Apparatus for applying chemicals in liquid form.

Sterigma, pl. *Sterigmata* Projection for supporting a spore.

Stipe A stalk.

Stipitate Stalked.

Strain An organism or group of organisms differing in origin or minor aspects from other organisms of same species or variety.

Stroma, pl. *Stromata* Mass of fungus hyphae often including host tissue containing or bearing spores.

Subiculum, *Subicle* Netlike woolly or crustlike growth of mycelium under fruit bodies.

Substrate The substance or object on which a saprophytic organism lives and from which it gets nourishment.

Suscept A living organism attacked by, or susceptible to, a given disease or pathogen; in many cases a more precise term than host but less familiar.

Susceptible Unresistant, permitting the attack of a pathogen.

Swarmspore Zoospore.

Synnema, pl. *Synnemata* Groups of hyphae sometimes joined together, generally upright and producing spores; coremium.

Systemic Term applied to disease in which single infection leads to general spread of the pathogen throughout the plant body; or to a chemical that acts through the vascular system.

Teleomorph State of life-cycle in which spores are formed after nuclear fission.

Teliospore Winter or resting form of rust spore, from which basidium is produced.

Telium Sorus producing teliospores.

Thallophyte One of the simpler plants, belonging to the algae, bacteria, fungi, slime molds, or lichens.

Thallus Vegetative body of a thallophyte.

Tolerant Capable of sustaining disease without serious injury or crop loss.

Toxin Poison formed by an organism.

Tylosi, pl. *Tyloses* Cell outgrowth into cavity of xylem vessel, plugging it.

Urediospore Summer spore of rusts; one-celled, verrucose.

Uredium Sorus producing urediospores.

Valsoid Having groups of perithecia with beaks pointing inward, or even parallel with surface, as in valsa.

Vector An agent, insect, man, etc., transmitting disease.

Vein-banding Symptom of virus disease in which regions along veins are darker green than the tissue between veins.

Verrucose With small rounded processes or warts.

Viroid The smallest known virus like infectious agent having no protein coat and only a small amount of nucleic acid.

Virulent Highly pathogenic; with strong capacity for causing disease.

Viruliferous Virus-carrying; term applied particularly to virus-laden insects.

Virus An obligate parasite capable of multiplying in certain hosts, ultramicroscopic, recognizable by the effects produced in infected hosts. Has nucleic acid with protein coat.

Wilt Loss of freshness or drooping of plants due to inadequate water supply or excessive transpiration; a vascular disease interfering with utilization of water.

Witches' Broom Disease symptom with abnormal brushlike development of many weak shoots.

Yellows Term applied to disease in which yellowing or chlorosis is a principal symptom.

Zoospore A swimming spore, swarmspore, capable of independent movement.

Zygomycetes Subclass of the Phycomycetes, characterized by gametes of equal size.

Zygospore Resting spore formed from the union of similar gametes.

Printed by Printforce, the Netherlands